ROLAND WOLDT

Successful Architecture Implementation

A practical guide on how to implement your EA/BPM program

First published by What's Your Baseline? 2025

First edition

ISBN: 978-1-967221-00-4

This book was professionally typeset on Reedsy.
Find out more at reedsy.com

For Tanja, Paul, and Carl — look, it just took 19 years to get it done.

Wake up, Neo ...
The Matrix has you ...
Follow the white rabbit.

— Trinity

Contents

Foreword

It Depends.

The two magic little words, like "I Love You", but for technology consulting. All at once, an answer, and a deflection, implying a depth of experience and a careful consideration of the inquirer's needs. It's the beginning of a conversation, demanding a further clarification of the business circumstances, testing the waters with proposals, before coming to a collective conclusion on a clear path forward.

I first witnessed the use of this phrase, expertly wielded like a scalpel, almost two decades ago, in my first Management Consulting gig. And since then, I've found myself repeating it, working to uncover challenges and propose solutions for some of the largest and most heavily regulated organizations in the world. I've found it immensely useful from the boardroom to the war room to the shop floor — it demonstrates humility, invites feedback, and gives you time to think.

It's also a lie.

The core of "It Depends" implies that the speaker doesn't know the right path forward. They need to uncover the specifics of your organization and its challenges, your architecture and technical debt, and your cultural reality, strategic direction and change management capabilities. It also tells you that your company is different from your peers, demanding a tailored solution to the problems you're uniquely faced with.

Truth #1: We often already know how we'd do it if this was our decision to make.

Truth #2: Your company isn't as unique as you think. Most of your problems can be addressed by similar solutions and structures already standard in your industry.

This is why "Successful Architecture Implementation" is such an important read for any business leader. I've been in countless meetings with Roland, and seen him in action, applying his deep expertise and repeatable approach to seemingly disparate organizational challenges. We've laid out the case over more than 100 hours of 'What's Your Baseline' podcasts, numerous webinars and celebrated conference presentations. This is the distilled knowledge — a blueprint for success, with actual answers, that can be immediately applied in your organization, or that can become a part of your consulting practice.

Beyond the solution itself, this book presents the case for an organization's awareness and focus on the discipline of architecture. Pay special attention to the 'why' of architecture — earning buy-in and sponsorship from stakeholders and executive leadership is *almost* as important as having the solution. As Roland loves to say, "It doesn't matter how much ink you put on the page — if the person reading it doesn't want it, it won't get done."

Go forth, good luck, stay strong, and we'll see you in the next one.

J-M Erlendson

Co-host, What's Your Baseline

Chief Evangelist, ARIS

General Manager, Countermeasure Music

Chairman of the Board, SING! Festivals

Preface

Imagine you just bought a hotel a year ago and spent all your time and effort on renovating it. Business starts to pick up and the first reviews from customers are stellar. You really pulled it off in transforming your hotel into a stylish destination, even though "the bones" of it are 50 years old and -luckily- you did not have to touch them to pass the code.

You are also having a small side wing with three conference rooms that have up-to-date audiovisual equipment, and you can host up to 80 people per conference room. Your "business wing" also has a small kitchen, and you can offer your guests refreshments and coffee. Life is good.

But now this happens — at 10:00 am on a Thursday, a potential client reaches out to you and asks if you can host a conference in your hotel three months from now. He expects about 600 to 800 participants, and the conference will take a full week. This is YOUR opportunity to get more revenue than planned for the first year in and also establish your hotel with the local business community.

But then ... you just have the three conference rooms and the small kitchen. What do you do?

- Can you just remove some interior walls and have some removable dividers instead of solid walls?
- Can you add wide opening glass doors on the exterior wall so that you have additional outdoor meeting space (assuming that the conference is in the warmer months of the year)?
- But what do you do in winter? Are you willing to lose business for half of the year when it is not warm enough for outdoor events?
- Will there be structural changes to be made — especially since you don't

have the original plans?
- How long will each option take?
- What will it cost?

Architecture to the rescue

Major changes are the situation that I see in many of my clients — they are starting a major transformation program, a system implementation, or a simple process or organizational change and "just do it". And the consultants tell them that they shall "fail fast and pivot quickly". Or IT implements some new technology without some form of user requirements/ analysis and wonders why the new app is not used or accepted (the equivalent of "hopping into your pickup truck, driving to the home improvement store for some 2×4s, and nails, and then starting to hammer everything together" in our scenario above).

Is that the way you want to manage your enterprise? I don't think so — and therefore this book is for you.

Architecture, which in my mind includes both: "Enterprise Architecture" (aka IT Architecture Management that is mostly done by IT) and "Business Architecture / Business Process Management" (the same exercise done by business folks). The main promise of any architecture work is fourfold:

- Capture the as-is situation (the plan of our hotel)
- Analyze the ideas and determine the impact on a future-state architecture (the options for change of the building)
- Planning and road mapping of the change to transform what you do to the next future state
- Monitoring and analysis if the changes have resulted in the desired business outcomes

While you are at it, you also bring in a new enterprise capability and tools. Which means you will have to interface with other programs and roll-out the "new stuff" to others in the organization, even though they might not be willing to change — so you have to convince them. And you also have to define the "rules of engagement" and determine how the architectural activities fit into existing processes — your software development lifecycle, your program planning, your (project) portfolio management.

Who is this book for?

This book is for the brave people who understand that organizations are complex systems and that changes in one place will have an impact on many people (ask anyone who has ever implemented an ERP system). Typically, every stone will be upturned in a transformation program.

It is for those people who realized that they have to implement an architecture capability in their organizations. They have to be able to describe, analyze, plan, support the implementation of changes, and measure their impact to be able to manage the enterprise in a responsible and sustainable way.

It is also for the people who have tried that before and failed. Maybe because they only focused on the implementation of a tool (the "make it, and they come" approach), or they "forgot" that people typically do not want to change.

It is also for the curious people who are not 100% convinced that this architecture thing is not too big of a bite for your organization. They would like to learn what it takes before they make the decision, if they are ready for it.

Whatever your situation is, this book has the answers for you. I will walk you through the major areas that you have to look at:

- An introduction into the relevant concepts of architecture — all the gobbledygook that you have heard before will start making sense.
- An approach to get your implementation off to a good start (the key idea of this book).
- Developing an architecture strategy that will allow you to not only identify the objectives and services that you will provide to your organization

in the future, but also to communicate it to the various stakeholders and measure your successes.

- Define the "rules of engagement" (governance) for your processes, the involved roles and org units, as well as the technology.
- The setup of architecture org units and the enablement of the various roles that are involved.
- We will also see how all of this will be applied to a slice of an architecture — the business architecture. You will see how you can set up this view, create/import content, and analyze it to be able to answer the questions from above.
- This book closes with an appendix that shows the foundations of the BPMN notation and data modeling to give you a head start on creating artifacts that we have discussed in the earlier chapters.

But why should you take the advice from me – aren't you just a software salesman who will promise me everything?

Well, you will find out :-)

But in all seriousness, I have been working in this field for almost 30 years and have stood up many architecture practices during this time. I also worked "in the trenches" for many years — as process designer/analyst, domain and solution architect, and as enterprise architect, advising my clients in the best way that met their needs.[1]

This brought me to work for half of the time at major consulting organizations (Accenture and KPMG), and the other half at the main software vendors in this field (IDS Scheer/Software AG and iGrafx). Over time, my responsibilities rose, and I led global org units like the global Center of Excellence for Process Mining, or the Global Consulting organization for a vendor. However, in all these roles, I have worked in the lighthouse programs for the clients who wanted to make significant changes and bring their organizations forward.

In addition to this, I am running the "What's Your Baseline?" podcast, in which my co-host and I are demystifying Enterprise Architecture and Business Process Management since 2021 and have interviewed dozens of experts who have freely shared their experiences and approaches. Give it a listen (watch) at whatsyourbaseline.com.

Oh, and regarding the software — you can do architecture without any software, using just whiteboards and (butcher) paper. However, you should use a proper tool for this, as we will discuss below. I am using a software called ARIS[2] as the example in this book — it is a model-based tool (vs. the more list-based portfolio tools that you typically see as "EA tools"), but that doesn't **really** matter for this book. It will make a difference in the configuration of the tool and the major audiences for the tool, but the concepts and approaches that you need to follow during your implementation and operation of your program stay the same.

So, now go and start reading. I promise it will be interesting and give you a new appreciation of the architecture discipline.

Acknowledgments

Countless people have already said that writing a book is a team sport. You need the feedback as an author to not only be held accountable for finishing the manuscript, but you also need others who will review it. These will not only "De-Germanize" your English and find typos, but also provide content suggestions, so that the final product is complete.[3]

But then you also require the people who have the experience of writing, publishing and marketing the book. And they also make you think about **WHY** you want to take on that effort in the first place.

I am happy that I have a team of people behind me who cover all of these aspects, and I like to thank them. So, without further ado (and in no specific order) here are the rock stars of this book:

- **J-M Erlendson**, who I have the pleasure of producing the What's Your Baseline? podcast with. Over the course of 14 years, first as one of my consultants, later as one of my peers, and now as a friend, he gives the necessary "reality check" on the topics in this book. Given the fact that he is also working in the media industry, I could learn a ton of audio and video editing from him — and he is an overall funny guy to hang with.
- **Randolph Schütte**, who I worked with on a multi-year project back in the 2000s with IDS Scheer, was the first sparring partner for writing a book ... which just took 19 years to evolve into this. He also was the first content reviewer and provided excellent suggestions, while confirming the structure and flow of the book.
- **Michael Schank**, who I was introduced to as a guest on the podcast and was the first real author that I met. He gave me some insights not only about the process of traditional publishing, but also the financial and

contractual arrangements that he went through with his book. This convinced me to go down the self-publishing route (what a rip-off traditional publishing is when you are the average Joe writing a non-fiction book in the area of your expertise).

- The **beta readers** of this book: Bart Gypen, Breanne Casteel, Caspar Jans, Dana Malcova, David Ruting, Ellen Pelletier, Eric Seidel, Gemma Hughes, Guillermo Lopez Felipe, Hanneke Loefs-Mos, Javed Qureshi, Marc Mendelson, Martin Holling, Marcus Zschuckelt, Matúš Mala, Ralf Gaumnitz, Ron Cohen, Tony Phillips — thanks for your interest in reading the manuscript and giving valuable feedback.
- All the **guests of the What's Your Baseline? podcast.**[4] Having over 100 recordings now out in the open showed me the vast variety of ideas and approaches to the topic of architecture — may it be business process management or IT architecture. They made this a better book by making me think if my ideas are really the best ones. I am proud to have adopted some concepts here, and I am sure that it is a better book than what I would have written just a few years ago.

And lastly: **YOU** — the reader of the book, the listener to the podcast, the follower on LinkedIn. I had so much fun and interesting conversations with strangers reaching out to me for a chat and then telling me that they feel that they know me (podcasting has the benefit that you are literally in the ears of your listeners while they drive to work, mow the lawn, do the dishes, or whatever else is the second thing they do while consuming the content).

Keep those requests coming. Reach out to others and grow your network. This will grow the discipline and its maturity immensely. And that is what we all should be aiming for.

I

The Case For Architecture

1

The Case For Architecture

Nobody wakes up one morning in their early years and says, "I want to become an [enterprise] architect!"

But there is a need for this skill set in organizations. The better ones — you know, the ones who seemingly can change on the fly — have implemented a practice ... either focusing on the whole enterprise, or in process management or IT architecture in a lower maturity state.

But what are some reasons for the sad state of affairs when it comes to architecture?

- Organizations have outdated or no documentation.
- Nobody understands what was implemented in the necessary detail.
- The impact of proposed or actual changes downstream is not clear and doing the analysis is a painful exercise, so often it will be replaced with a "wild guess".
- You will be able to collaborate with all affected parties. There might be a tension between business and IT as the two largest players here, that can be resolved by creating a shared view of the changes.
- Upstream planning ("How does my change fit into the bigger picture, or our project landscape?") is also a wild guess, and the pet project of a stakeholder gets priority.
- The potential improvements of your suggestions are not seen or cannot

be quantified. Creating a business case becomes a nightmare (of wild guesses).

- Regulators might require the establishment of the capability, and you have to stand it up if you like it or not. But you don't know how to do this...

If you go down the route of implementing an architecture program and capability, you can reap multiple benefits. Some of them are:

- Enable planning and 'bigger picture' when it comes to structures, architectures, and projects.
- Establish common language and approach in your organization.
- The structured discipline will help to find operational and compliance improvement potentials. You can literally see what is going on in your organization, and only what you can see can be improved!

What's in this section?

This section sets the stage for you and why you want to establish this capability in your organization as well. We are looking at:

- The changing role of architects.
- Which different levels of capabilities you need to build in your practice.
- The different inputs that confuse many people and make them think that architecture is a nerdy and niche topic, while to the opposite they can accelerate your efforts (frameworks, notations, methods, and reference content).
- The Solution/Process Lifecycle as the structure in which all architecture activities take place.

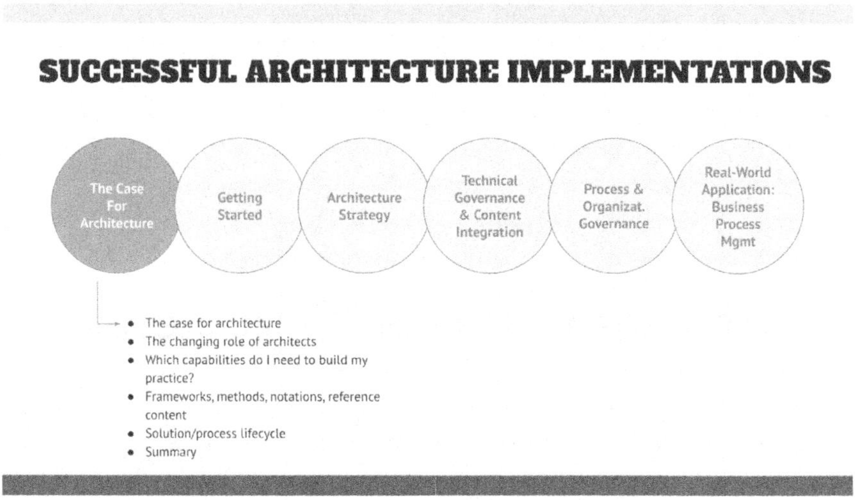

Fig 1.1: The case for architecture structure

2

The Changing Role Of Architects

Even though the discipline of architecture[5] is decades old (and if you look at the discipline of process management, even hundreds of years), there is no consensus of what an architect actually defines. And the practitioners have not been helpful as well in their self-positioning.

"Back in the day" a lot of architecture groups held one type of architect – the people defining the methods, tools, and standards. The folks who created "the committee of no"[6] (aka the Architecture Review Board), who were excited about frameworks, notations, and all those other things that don't have an immediate impact on what was considered important for business people. That led to many occasions where architects have been excluded from projects and meaningful conversations/decision-making. Even though, they should be the people at the table who understand how things are wired up and give good input when defining the right solution for a business problem.

Today, however, there is a change of the role required: for once, the old role description has not worked out, and the architecture content has become more complex (think about the aspect of "co-opetition" with your former competitors is requiring to share forces at times, while competing in other situations – that has not happened before). Or on a technical level, solutions have become smaller, and smaller down to a "function as a service" approach in the cloud, which is vastly different from the "in-house, we standardize everything on our ERP system" approach from 10 to 20 years ago.

This leads to changes in the roles of architects that not only cover the skills of the individual (for example, you did not need to be an outsourcing/cloud contract specialist a few years ago), but also the positioning of what it means to be an architect.

Archetypes of architects

But how do you define the various archetypes of architects, then? A practical approach is to look at it from two perspectives:

- The **scope of contribution** (from "low" to "high" as in defining the standards but not actively contributing to the solution solving).
- The **degree of specialization** from generalist to specialist.

Figure 2.1 below shows the four archetypes in a 2×2 matrix, and will provide some insight in why you hear different answers if someone asks for an architect to help with implementing a solution.

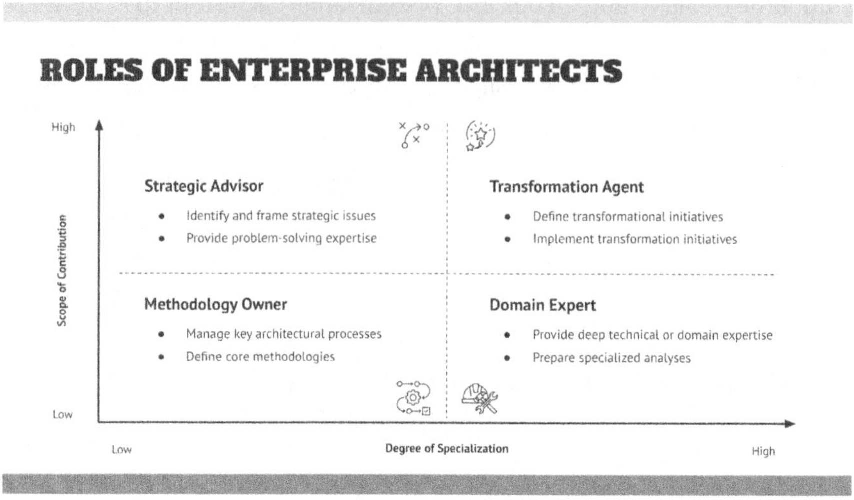

Fig. 2.1: Roles of Enterprise Architects

- In the low/generic corner you find the **Methodology Owner**, the traditional role of EAs described above. This is a comfortable place where you can define the core methodology (while not being asked to develop the artifacts) and manage the architectural process.
  This could be one reason for the popularity of this position – you are "important" without having skin in the game – no wonder why this does not resonate well. However, even in the future, defining the governance and "rules of engagement" are still critical when setting up your architecture implementation and running your architecture program.
- In the low/specialized corner, you find the **Domain Expert**. These are the people who will describe themselves as "Business Architect" or "Data Architect", or sometimes even more specialized as "Salesforce Architect" or "Security Architect" if they bring some very specialized knowledge to the table.
  This role provides deep technical or domain expertise and creates artifacts or provides detailed analysis for their area of expertise. Un-

fortunately, the term "architect" is not well specified, and sometimes people call themselves architects just because it sounds impressive, but have no clue about architecture concepts, frameworks, practices, etc.

- The high/generic quadrant holds the prestigious role of **Strategic Advisor**. This is the "seat at the table" role that some architects think they should hold where they can help define the "what" by identifying and framing strategic issues, as well as providing problem-solving expertise.
- The last quadrant holds the **Transformation Agent** (high/specialized), who is focused on the "how" to implement the "what". This typically includes defining and implementing transformational initiatives, and supporting the management of change in general.[7]

As one can see, the two high contribution roles of Strategic Advisor and Transformation Agent build the bridge between strategy and Project Management Office (PMO). This led to the analyst's prediction that the Strategy, Architecture, and PMO groups would merge into one group. Unfortunately, I have not seen this anywhere yet.

Figure 2.2 below shows the mapping of activities (and the amount/volume of work shown as bubble size) for one client as an example. It becomes clear that this group has successfully extended their sphere of influence to the strategic quadrant, but is short on the "how" of implementing changes.

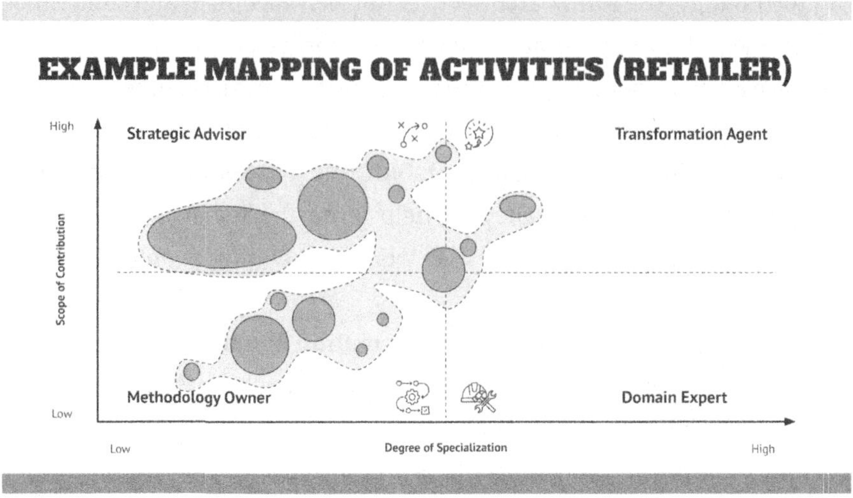

Fig. 2.2: Mapping of activities for a retail client

When I look into my personal "crystal ball," I see the role of Methodology Owner being further reduced to a minimum level (thanks, Agile). The role of Domain Expert will further splinter into smaller and smaller specialties before a readjustment to the main domains (strategy, business, data, apps, technology) will take place. Roles like "app xyz architect" will become less prominent.

On the strategic/transformation side, I see some progress, but maybe not in the way that the analysts see it (the elevation of the EA group to the table). This would require that the discipline further matures and an agreement on how to measure EA activities would be defined as an industry consensus.

Until then, it will be individuals, some of them former architects, who will get into the driver's seat to define the "what" and "how".

Roles in an architecture program

So, how do these changed architecture roles now find a way into a program? My recommendation is to look at the various levels of a program and determine the degree of specialization and contribution:

- **Solution level**: this is the project that delivers an implementation – a reorganization, a process improvement, or similar. The architecture roles on this level are the **Domain Architect** or **Specialist** – people that are mentioned above and hold fancy titles like "xyz architect". These folks are the ones creating the artifacts and designs, performing the analysis, and coming up with recommendations in their field of expertise.
  However, when implementing a complex solution, you need to have someone who brings these individual results together. They make sure that the product of the project is complete, usable, does not have gaps, and does not overemphasize on one area of expertise while lacking quality in the others. The **Solution Architect** also coordinates with the project/program level and has a semi-role of project manager for the architecture development.
  Furthermore, please note that this role is not included in Scaled Agile,[8] which, I think, is a big gap in that framework.
- **Solution Portfolio** level (aka "large solution" in Scaled Agile): here you find the role of **Solution Portfolio Manager**, who is responsible for coordinating the various work streams and defining the runways that need to be built before a development can be done.
- On the **Enterprise level,** you find the **Enterprise Architect**, whose main task is to start and maintain the system in which architecture development happens. This includes the governance areas, as well as the tool, support, and knowledge management topics.
  This role typically will be supported in the day-to-day work by a Center of Excellence. It typically consists of various supporting roles, such as Developers (in the EA platform, creating scripts and dashboards), Administrators of the central EA platform, and Power Users who will help the teams with methodological support regarding specialized analysis situations. Think Simulation, Cost Analysis, or Process Mining as applicable use cases. They can also be the first level of support for end users.

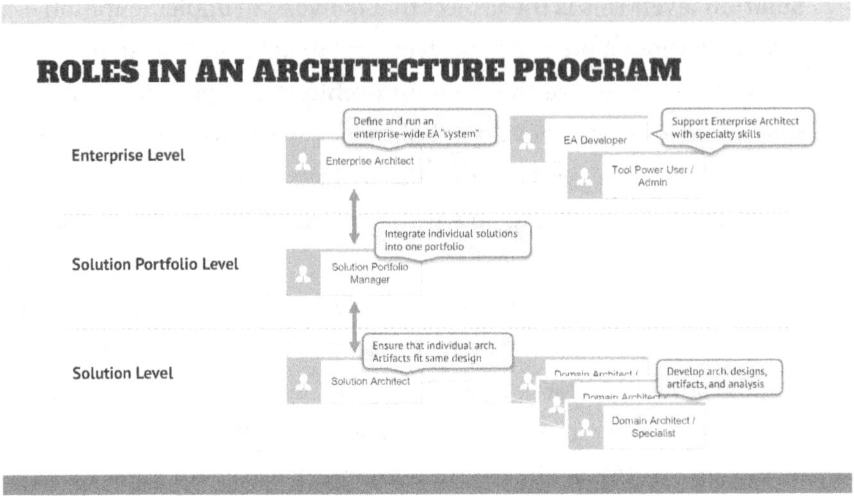

Fig. 2.3: Roles in an architecture program

Obviously, these roles need to be embedded into organizational structures that are defined as part of the EA implementation. Figure 2.4 below shows the application of the roles in a federated governance model:

- First, you see the **individual projects** (small circles) within the Enterprise, that develop the architectural artifacts by domain experts and put together into a consistent deliverable by the Solution Architect.
- You also see how the Solution Portfolio Manager is coordinating the various developments in his/her **architecture segment** (larger circles). For now, business as usual as described above.
- What you also see is that each project interacts with the **Architecture Review Board**; that is one of the governance levels (the judicial branch of government)[9] that we discussed in an episode of our podcast.[10]
- Lastly, you see the **EA Center of Excellence** (COE – the "executive branch") as the service organization that provides its services to all participants in that system, and ensures that things run smoothly.

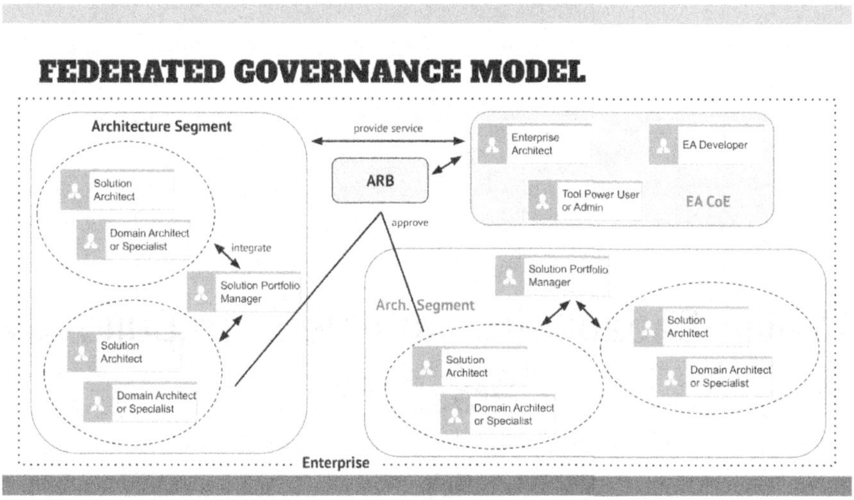

Fig. 2.4: Example of a federated governance model

If your organization currently follows a more hierarchical governance model, you can still use the role definitions above, but your approval and governance organizations will be different, obviously.

3

Which Capabilities Do I Need To Build My Practice?

Defining an architecture capability model appears to be difficult because the term "Enterprise Architecture" is such a loosely defined term, which has changed over the last several decades.[11] This leads to confusion if EA is "just the tech stuff that I don't understand" or if business people rather look for BPM (Business Process Management) instead.

This chapter sets the foundation for the more how-to-focused approach of chapter 8 below. The objective is to establish the basis for services that the practice will provide, or how to grow the maturity of the practice and the practitioners.

Three areas of a successful architecture implementation

One takeaway of dozens of EA implementations is that three major areas determine a successful implementation:

- the content that architects create,
- the setup of a proper governance, and
- the adoption capabilities.

Please note, that the technical implementation ("getting the tool to users") does not show up as a key capability, which is something that some people find interesting and counterintuitive.

When you look at the "three-legged stool" that you need to balance for your organization as the implementer, we see three areas: Content, Governance, and Adoption.

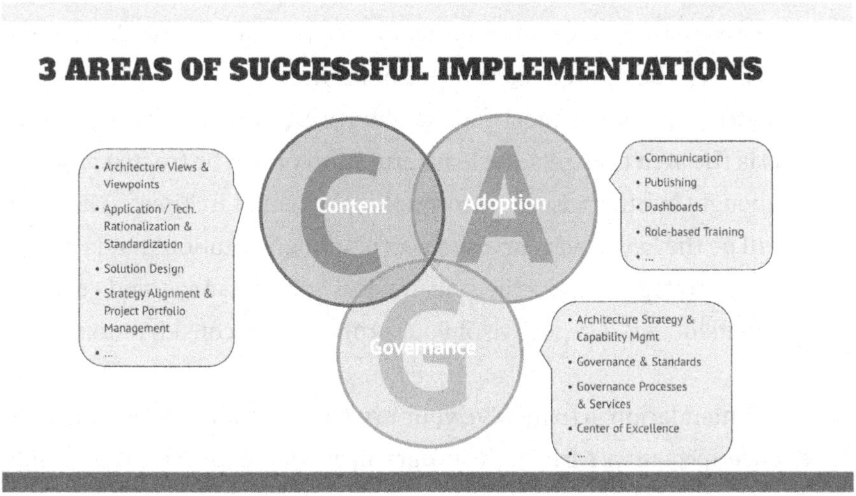

Fig. 3.1: Three areas of successful architecture implementations

- **Content**: this is most likely the area that brought you to identifying architecture as a key success factor for your program and got you started. It is the problem that you want to solve, and typically, you see topics like the ones mentioned in the graphic above (figure 3.1). Additional topics might be Process Mining, Robotics Process Automation, or Simulation. In this group, you will find the most advocates of the 'give them the tool and get out of the way' or 'the tool is just Visio on steroids' approaches. Don't adopt this mentality – it will lead to a low adoption rate and ultimately make your program fail.

- **Governance**: this area defines the direction and 'rules of engagement' for the program and are already mentioned in the related chapter. In a nutshell – you must define what your participants shall do (Process Governance), how they are organized (Organizational Governance), and which standards shall be followed (Technical Governance). This area describes the 'inside-out' perspective – how do we get things done? In the past, many EA groups have seen this as their main focus – maybe because they could not discuss the content topics? — which led to the impression that they built their ivory towers (or were the 'Committee of No'[12] vs. the desired 'Committee of Know').
- **Adoption**: As in the most failing enterprise-wide implementations, this is the area that gets the least attention or will be treated as an afterthought by otherwise more technology-oriented architects. However, it will be the key driver for a successful implementation that is driven by your strategy. It takes the stakeholders, user personas, and -overall- the 'outside-in' perspective of your program into consideration.

My recommendation is to involve your org change management and user experience specialists early in the program, and let them drive the outside view, while others take care of the governance and technical aspects.

Depending on your current perspective or situation, you might focus on one area over another, but keep in mind that you want to balance these. During your implementation, you might have to shift the focus according to where you are on your journey. Initially, there will be more focus on governance and adoption, while content will become more in focus once all participants in your program have understood what is expected from them.

What are capabilities?

First, let's talk about what capabilities are not: they are not skills and knowledge that an individual has. They are also not features and functionalities that a software has.

Therefore, it is good to make a distinction between **business capabilities** and **technical capabilities**. The latter are things that an application can do and fits more the second scenario above. In earlier versions of my favorite architecture tool, this object was called "IT Function", and I think this is a better term than (technical) capability. It makes it even clearer in the context of an application rationalization, where you want to find redundant functionalities to streamline your application portfolio and avoid multiple apps that do the same. You might want to find missing functionality that you can provide with another app/extension or build by yourself.

The more interesting term, in my opinion, is the business capability. This term means the definition of the "what" an organization shall be able to do. It consists of things shown in the middle of figure 3.2 below — organizations, skills, knowledge, locations, applications, processes, data, and more. Together, these items provide a good high-level description of what shall be implemented, updated or retired.

Business capabilities are a core concept of every architecture – they are derived from the strategy (which is not only comprised of the items below, but also includes the services and products, the business model, and other things), and should align to it. The business capabilities are then driving the programs on the roadmap and can be used to be a litmus test for each project request by asking "which capabilities do you create/update/retire ?". If the answer is "none of those," then the justification for that project funding request is on a weak foundation. In an ideal state, everything should be aligned with the capabilities, and the results of the projects should be measured and show the accomplishments of the strategic objectives.

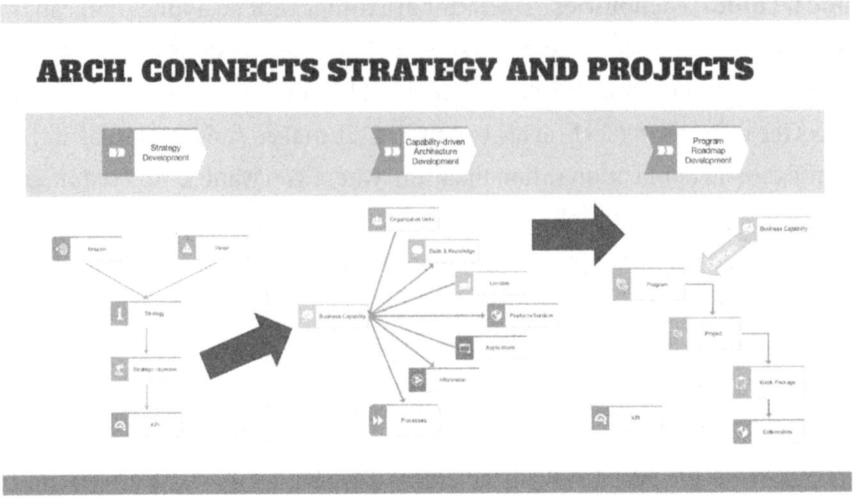

Fig. 3.2: Capabilities bridge strategy and projects

Implementing projects require a bit more detail, though. When you look at your organization, you will find different states of maturity, or needs and requirements which might be a result of your organizational structure (some org units might have more people than others, and therefore might have the need for more automation and workflow, while the smaller regions might be able to get away with a more manual approach).

To show this and have objects available in your planning and alignment during the architecture development, the concept of *capability configurations* comes into play. Capability configurations are subsets of the main business capabilities and might be defined either as different fidelity states of a capability (think "gold/silver/bronze" in the example above), or various states of the capability being laid out on a timeline. Either way, the capability configuration is what needs to be defined to describe the scope of a project (and, of course, shows the specific processes, affected apps, data, etc. in more detail).

An architect will have to integrate this back into the overall capability model, answering the question of "how far are we in implementing the overall capability model?". This will then provide the feedback to the strategists about what the organization might be able to do soon, and what the impact on products/services or program delivery might be.

Architecture capability model in more detail

So, what does that mean for your architecture practice? In my experience, they are grouped into five big buckets of business capabilities, shown in figure 3.3 below:

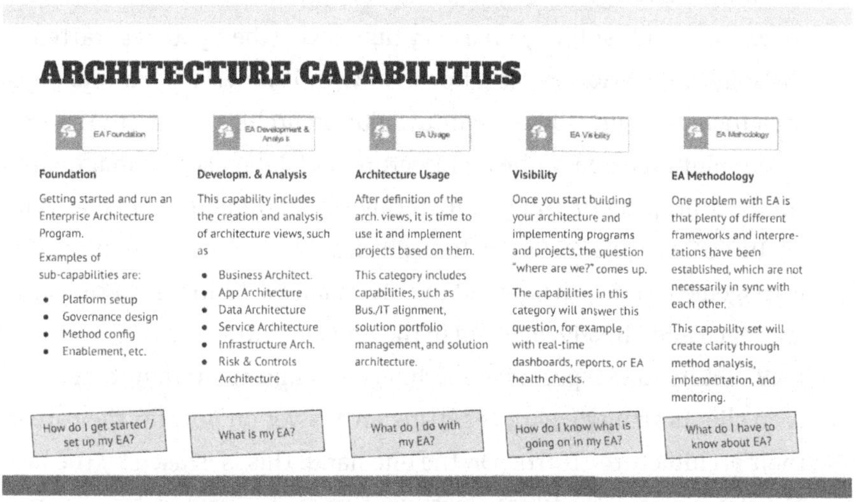

Fig. 3.3: EA capability model on a high level

- **EA Foundation**: This bucket includes all capabilities that are needed to answer the question of how to set up and maintain an architecture program. This includes the technical aspects of your chosen platform components, and governance, but also the people side of things – the governance and adoption circles in the Venn diagram of figure 3.1 above, as well as the high-level content structure (which framework to follow, the structure of the meta model,[13] and so on).

- **Architecture Development and Analysis**: this is the first area of content, and it includes the capabilities that cover the various views of your architecture. If you are following the TOGAF framework, for example, this covers the Architecture Development Method phases A-D. Besides modeling the content in Enterprise Baseline to get an understanding of the as-is ("what runs?"), this capability bucket also includes the various analysis techniques, such as application rationalization, technology standardization, or process mining. It basically aims to answer the question "what is my EA?".
- **Architecture Usage**: Once you understand what your as-is architecture looks like — most likely on a very high level when you get started — the logical question is "what do I do with it now?". This capability area includes things like creating solution architectures (in the Projects/Solutions area of the architecture tool database), managing the solution/project portfolio, or fostering the alignment of business and IT. In TOGAF, this would include Phases E to G and the requirement management (Phase H is part of the governance definitions in the overall approach and already included in the EA Foundation area).
- **EA Visibility**: one aspect of the architecture implementations that I have seen clients struggle with is getting an answer on how far they are in their architecture efforts. On the one hand, this is caused by the lack of a strategy, and therefore objectives that are not well-defined, which then leads to arbitrary KPIs that will be measured. On the other hand, this directs the pure focus on non-capability-oriented measures, or "formal" measures. An example for such a KPI would be how many training sessions have been delivered as part of a go-live. This is interesting information but does not tell the story if the implemented solution brings in the business results that should be achieved. That would be better measured six months after go-live when looking at the business results.

  In addition to this, many programs did not do their stakeholder needs analysis homework and determined what the information needs of these stakeholders are. This leads to generic dashboards that are mostly not relevant to an individual stakeholder and want to do too much (but are easier to build).

- **EA Methodology**: when one gets started with the topic of architecture, there is a big learning effort in basically understanding the strange language that architects use, which also contributes to the "ivory tower" perception of them. However, many EA concepts and approaches are very valuable and therefore the question in this capability bucket is "What do I have to know about EA?". This covers things like which frameworks or notations to use, but also assessing if a new concept comes up (for example: "Do I still require an architect when I do Agile?"), and if the answer to these questions is that it is, in fact, relevant, then you need to plan how to implement this (and the impact on existing work), and actually make the changes.

One suggestion is to capture all the architecture capabilities needed in a multi-level architecture capability model. This model can act as the catalog for all the things that you need to define and implement to get your architecture program up and running — nicely laid out on a timeline later. The capability model that I brought to clients as reference content in the past included the five major capability buckets above, which were then decomposed into 24 level 2 capabilities that included ~110 level 3 capabilities.

EXAMPLE: EA CAPABILITY MODEL

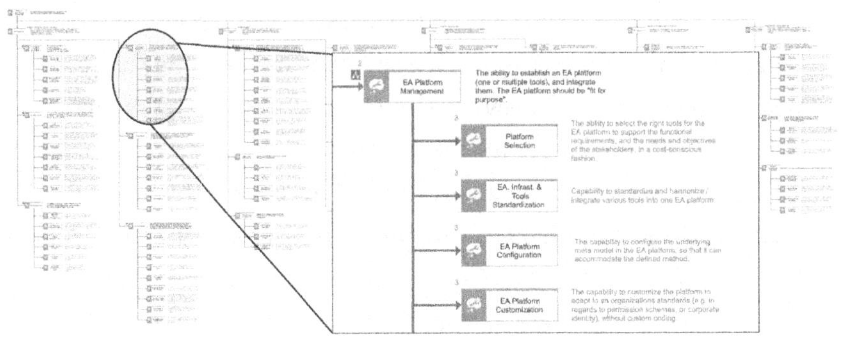

Fig. 3.4: Example EA capability model

My suggestion when creating an architecture capability model is to limit yourself to 3–4 levels at maximum so that you don't lose the high-level picture and get lost in the details (which are defined in your cap configs anyway) – this is the time for broad strokes.

I further recommend using the capability model to determine the services that the architecture organization should provide when developing an architecture strategy. It can also be used for assessing the architecture maturity, which will also be a topic for the future.

4

Frameworks, Methods, Notations, Reference Content – Some Clarifications

When speaking with clients and prospects who are new to the topic of EA or BPM, a lot of the key terms, such as Enterprise Architecture Frameworks, not only seem to be strange, but are also used interchangeably. This chapter will take the four most used terms, define them, and will show how they are relevant to getting started with your architecture implementation program.

Enterprise architecture frameworks

This is the one term that is marketed as the "special sauce" by enterprise architecture framework providers (and they are happy to sell you training, certifications, and even license it to you), but in all reality they are considered buzzwords and don't have a day-to-day relevance for practitioners.

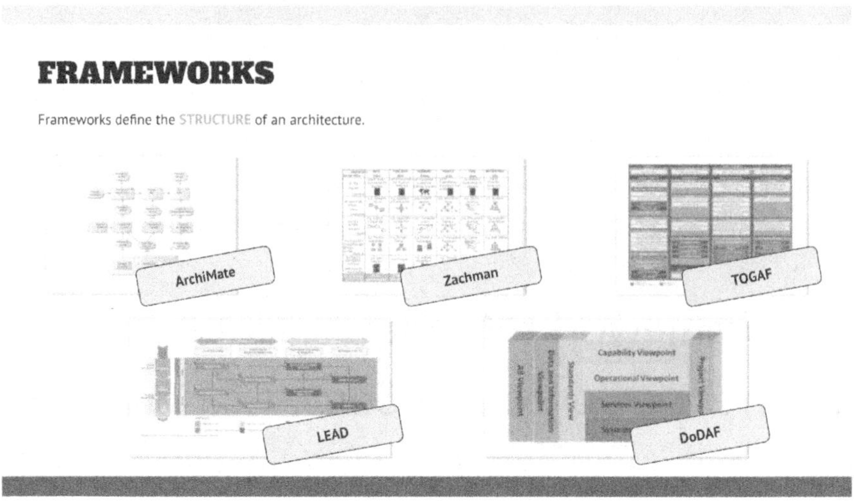

Fig 4.1: Frameworks are structures

But what are EA frameworks?

Starting with the Zachman framework in 1987, through Professor Scheer's Architecture of Integrated Information Systems (ARIS) framework, and The Open Group Architecture Framework (TOGAF), practitioners in the EA space have developed plenty of different frameworks. They mostly want to help with structuring the content in a way that will help users to create content repeatably, and provide guidance on where to store and find this content, typically defining views and viewpoints for the content.

In addition to this, some enterprise architecture frameworks also provide a method (for example TOGAF) or a notation (for example ArchiMate), while others exclude these extensions and focus on being more specific on the individual artifacts (the "what") that shall be created. However, the Department of Defense Architecture Framework (DoDAF) explicitly does not describe how things shall be depicted, but rather follows the mantra of fit for purpose.

This makes comparisons difficult and blurs the lines between the various frameworks, while on the other hand, some industries require following a specific framework to get approvals. In the US, for example, not only the DoD requires the use of DoDAF, but other federal and state/local governments make it mandatory to use DoDAF when implementing systems.

Why are they relevant, and how should you use them?

As said above, the main benefit of selecting an EA framework to follow is the alignment of users and the expectations of where to find content that they are looking for. The overall benefit is the reduction of time for searching content, and training effort.

This will be implemented in the folder structure that, for example, has two folders (Enterprise Baseline -"what runs?"- and Projects/Solutions – "what changes?") on the main level, while the structure below will be determined by the various views that the framework prescribes. When a project goes live, the designs that are in "Projects/Solutions" now can be moved into the Enterprise Baseline to reflect the new "what runs", and be available for future analysis and reporting.

Methods

Methods are processes that describe how things are done. These can be related to a framework (see TOGAF), or be available as a standalone artifact. Another example might be the Scaled Agile Framework (SAFe)[14] that not only contains the structures for scaling Agile (e.g., the "Trains"), but also includes the steps that need to be taken at each level of the concept.

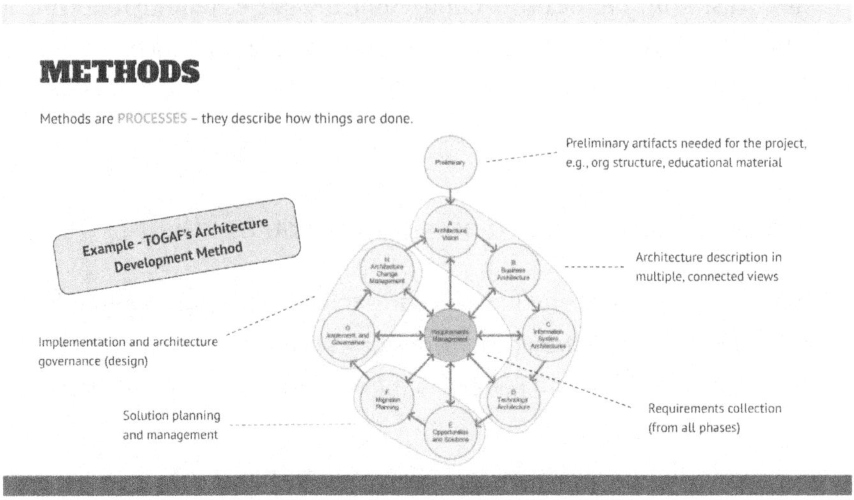

Fig. 4.2: Methods are processes

What methods are available?

There is a myriad of methods out there in various degrees of sophistication. Some of them have become the "classics" and will be taught at universities, such as TOGAF. Others are proprietary, and copyrighted by a consulting firm or an industry consortium so that you can use them only when you hire that organization for a project. Some others even claim to be open source, while their terms and conditions are in complete opposition to the four freedoms of free software.[15] They just want to have community input that then will be included into the content that is available for their paying members.

While your requirements for methods might differ greatly, I recommend doing some research and looking for industry-specific methods, and ask your tool or consulting partner for recommendations. Groups on LinkedIn and other forums might also be a good place to ask for recommendations.

Once you have narrowed down what method package might be applicable for you, I suggest you look at the provider and the maturity of the method content. Check if it is well documented, if training is available, or if an ecosystem exists around this methodology. Lastly, I would review the material together with your fellow architects and determine which experiences and favorites already might exist in your organization.

Why are methods relevant, and how do you adapt them?

Methods can be very helpful when designing architecture artifacts, especially when you are beginning with this exercise. However, almost all methodologies need to be adapted to your organization's needs. They might be very extensive (the TOGAF spec is 900+ pages, and very "dry"), or they might have an "old-fashioned" flavor to them – to stick with TOGAF again: the deliverables in the spec have a document focus, such as the Architecture Description Document, while your design in an architecture tool will be much more flexible and "living" than a document on (virtual) paper that nobody reads.

But the content that is described in the TOGAF spec is relevant and will ensure that you don't forget important things. However, you just have to adapt the content to your setup, processes, and tools. My suggestion is that you set up a methodology group that defines your organization's standards for architecture work. This will cover the artifacts that you will create, the configuration of your tool (see structure above and the notations section below), and cover the EA Method capability.

When you implement the standards, some things might seem strange to your audience, so encourage them to follow the rituals of the method. For example, the stand-ups in Agile feel weird when you get started, and so might be the fact that you don't produce documents as designs, even though you maybe have done this in your organization for years and decades. My suggestion is to give those unfamiliar and uncomfortable things a try, and trust the method before you make changes and miss the benefits – remember: people inherently do not like to change, and people will be the biggest roadblock on your implementation road to convince and bring onboard.

The outcome should be a set of documented standards that help your users to "ease into" – don't start with a 400-page document so that your adoption will not be hindered by this. Include your standards into the training material for new users and provide support offerings for them (an internal forum, an email inbox for suggestions and questions, "open hours" for real-time interaction with your users). Also make them available to act as a reference and tiebreaker when decisions need to be made, and treat them as a product that follows its release cycle and gets updated regularly.

Notations

Notations define the graphical representations and "grammar" for the artifacts that are created – the language/syntax. They prescribe the model types that shall be used for a given topic (e.g., lower-level processes or data models), and the objects that are used within those. In addition to this, the notations define rules which objects can be connected to which, and which attributes need to be used.

This offers the benefit that database-driven architecture tools can check the completeness and correctness of models before submitting them for approval or implementation, so that missing mandatory attributes or incorrect modeling (preventing automation) can be detected. Costly rework during the implementation can be avoided. It will also reduce implementation risk because the number of changes during coding compared to the blueprint will be reduced.

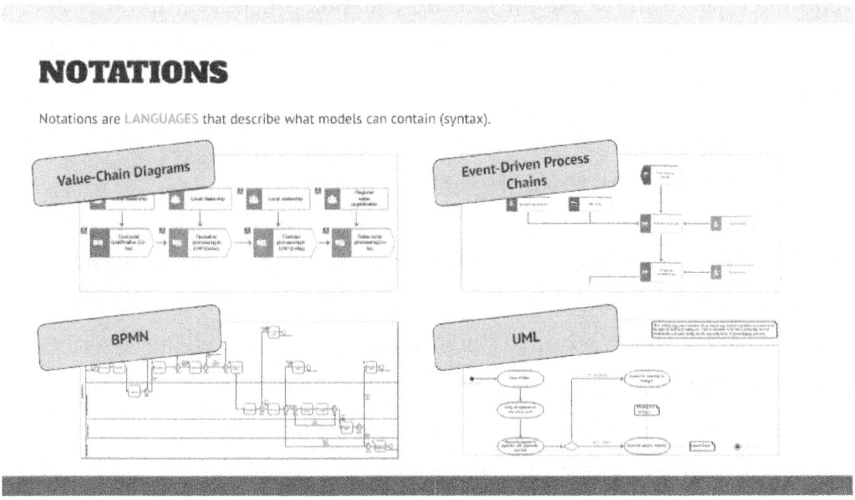

Fig. 4.3: Notations are languages

Which notations should I use?

This is a very loaded question, and you will hear a ton of different answers when you ask others. There are industry standards out there (BPMN,[16] UML,[17] DMN[18] to mention a few), but the correct answer is "it depends". Take a look at what you want to describe, and in which view the artifact lives (see the structure topic above). It also might help what your chosen tool recommends – which might not be very helpful when ARIS, for example, ships with 120+ model types out of the box. However, it also ships with a pretty good method help that will guide you, and there is an ecosystem available where you can ask more in-depth questions (forums, professional /consulting services).

For processes, for example, some generally accepted conventions have evolved: a few levels of Value Chain Diagrams and one level of either BPMN or EPC processes below. Which of the two latter notations you might want to pick is a matter of personal preference, ease of adoption, and tool capabilities (some features might require a certain notation – for example BPMN is required if you want to use the out-of-the-box process compliance check in ARIS Process Mining, while the synchronization with SAP will work with either notation).

How do I implement notations?

The implementation of a notation has multiple aspects: one is the selection of notations as described above, the other is the implementation in your tool, which is part of the "Technical Governance" phase and will be described in more detail later.

The next step is to create training and support material, and communicate the benefits of using a standard language, especially if it is an industry standard notation: your stakeholders and users will learn skills that can be reused in future projects or at other employers. It will also simplify communications with other parties (other projects, business partners, vendors) significantly.

Lastly, the standards group mentioned above should also solicit feedback in updating the use of notations in the field (what works/what doesn't work?). It also should look out for changes in the specifications and the use of notations in the industry, so that they can be incorporated as changes. However, switching notations completely should be a very rare event because it might influence rework, analysis, etc. In most cases, selecting a notation (vs. allowing "free form modeling") and sticking with it is enough – you should have a good reason for notation replacements.

Reference content

Reference content is pre-packaged content from a vendor or other organizations, such as process models on all levels of the hierarchy. It aims to accelerate project delivery, and to implement standards and patterns that will then simplify the selection of technologies, or leading practices as a secondary benefit.

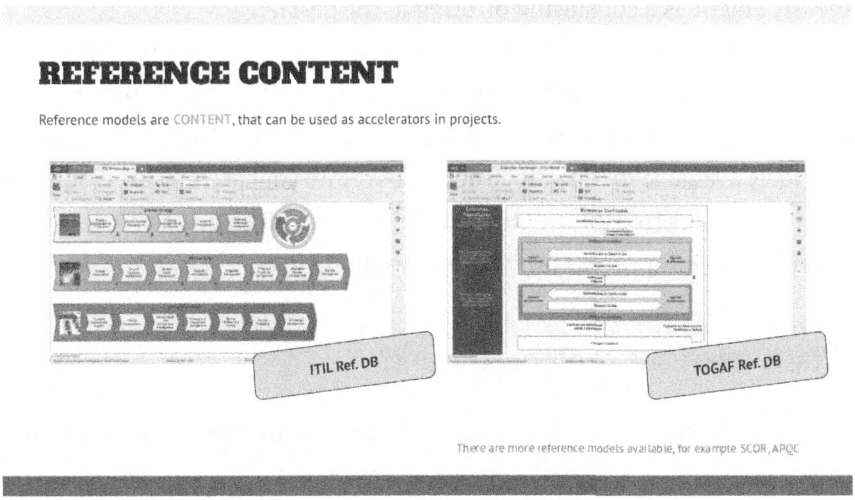

Fig. 4.4: Reference models are content

What is reference content?

Reference content can come in different formats – from Excel sheets, to databases in a tool vendor-specific format and proprietary tools that are built by the reference content provider.

An example of reference content is the Process Classification Framework (PCF)[19] by the American Productivity and Quality Council (APQC). The PCF is a decomposition of process titles on 5 levels that are structured in 13 main process areas (the "what"). It includes not only a generic PCF, but also industry-specific versions, which will make things obviously more interesting for customers. What it does not bring to the table is the content of the lower-level processes (the "how") because this is determined by too many factors, such as the selection of tools, or the org structure of an organization. This type of content might be available from a tool vendor (SAP, for example) or a consulting organization (Software AG, for example, offered in their Industry.Performance.Ready content that not only included process models, but also a pre-configured SAP system and performance measurement as a consulting offering in the past).

In addition to the process classification, APQC also adds definitions of typically used operational KPIs to the package, and offers industry benchmark values of these KPIs as an add-on service. The PCF and the KPI definitions are freely available for download and import into your architecture tool, and APQC also offers free information about benchmarking, and how to create a baseline for your process improvement activities.

Why is reference content relevant?

Very often clients ask for the creation of reference architectures and looking for an industry-specific reference content to be used as a baseline can be a considerable accelerator. In addition to this, your proposed content will gain more credibility if it is based on an acknowledged provider (e.g., the tool vendor, your consulting partner), or an industry consortium. Enabling comparisons, as described in the APQC example above, is also a welcome benefit.

In some cases, e.g., with SAP, the reference content is also enhanced with additional information that enables the synchronization with an architecture tool that is used for developing the blueprint in the Solution Design phase of the project. The large consulting firms also have developed their own reference content as part of their offerings, which might be based on other reference content (APQC is a favorite here). It is enhanced by their project experience and the leading practices that they've seen and helped clients to implement.

In addition to this, collecting reusable patterns that are offered as reference content (or developed by yourself in your architecture tool) will allow projects to design their solutions faster. If they adhere to the reference structures, an approval by an Architecture Review Board, for example, might be "rubber-stamped" because the due diligence for solution components might have been done already (for example a security review of a cloud provider's offerings).

5

Solution/Process Lifecycle — Architecture Is More Than "Draw It And Forget It"

Some organizations and also methods, such as Lean or Six Sigma, have a focus on improving a single process with a well-defined scope, while others take a wider, more holistic approach to processes and process improvement. This has an impact on how things are approached, where to spend time and focus on, and to which level of detail certain steps will be done in the solution/process lifecycle.

As always, you can discuss the differences between these approaches and their benefits. This article introduces the "wider" perspective of the solution/process lifecycle, which applies to process improvement projects, system implementations, or any other project type where architecture will play a part in.

Two areas — design and execution

The first thing that one needs when looking at the solution/process lifecycle is to distinguish the fact that there are different tools for different purposes, but in general, they fall into two categories:

- **Design-time tools**: these tools allow you to create blueprints, or analyze existing content. The artifacts in this area will also be re-used to compare the long-term execution with the initial objectives. Classic architecture tools, such as ARIS or HOPEX, fall into this category.

- **Runtime tools**: these are the actual transactional systems and support systems that allow the monitoring of them. SAP, Salesforce, Dynamics, Workday, and other systems fall into this category. Depending on the capabilities of the actual runtime systems, you might have to complement them with BI tools for monitoring.

SOLUTION LIFECYCLE

Fig. 5.1: Two areas for processes

The one mistake that I see clients make is to value one side over the other. Of course, you cannot run a transaction from a modeling tool. But without a proper design that can be critiqued by others (so that it shows the implementation design and plan in a granularity that there are no misunderstandings when it comes to configuration, coding, and testing) the implementation is already falling behind. On the other hand, many architecture-focused teams tend to be too detailed and strict about what needs to be created for approval. The implementation gets delayed unnecessarily and people perceive architects as a roadblock instead of a valued contributor.

Phase 1: Strategy and the Enterprise Baseline

In a perfect world, one would start with a clean slate and start creating the repository of what is currently in production in all architecture views as defined by your chosen framework (such as TOGAF).

In this phase the structure will be set up, the data and meta models being defined as part of the technical governance phase of the implementation, and existing content will be imported and normalized. *The main idea here is to create the "Enterprise Baseline" as the always current version of "what runs".*

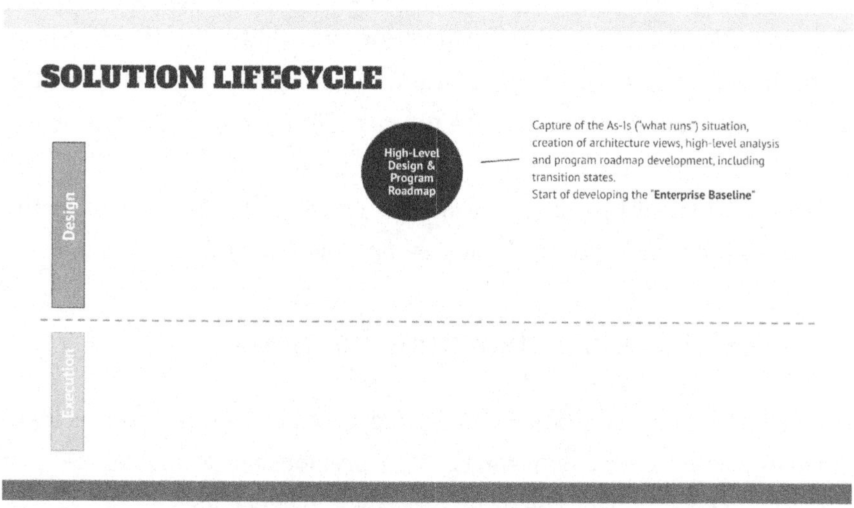

Fig. 5.2: Strategy and Enterprise Baseline phase

The good news is that you don't have to create everything at once, which would be a Herculean task that would also reinforce the impression of the "ivory tower architects."[20] The way to avoid this scenario is to provide "just enough architecture" guidance (principles and a minimum of reference architecture content, such as preferred technology). Work with your current projects on the definition of a minimum architecture structure (formal and content-wise). Bring in their designs into the repository – in the "Projects/Solutions" area, and in the case of approved designs, or projects that are close to go-live/just went live into the "Enterprise Baseline" area.

You can also use modern technologies like Process Mining and Task Mining (Robotics Process Discovery) to identify current state content and create straw man models to complete an initial picture of the architecture. You should also think about importing the current strategy artifacts so that you have a baseline set of objectives, requirements, needs, KPIs, etc. that you can map your (future) projects against.

The main purpose of the Enterprise Baseline is to provide the foundation for the identification of change and improvement needs, for example to reduce technical debt, to enable process improvements, or to define the structure changes of the organization. In addition to this, analysis techniques, such as process simulation, can be applied and help with the identification and justification for a future project (to support the business case).

These identified projects then will be created as placeholders in the "Projects/Solutions" area until they are approved and further developed.

Phase 2: Solution architecture and design

In this phase of the solution/process lifecycle, the solution that was identified as a potential problem solver in the previous phase and defined as a project will be designed. The result describes "what changes".

A "solution" in this context is more than the software that is chosen, but it is the summary of all views (processes, apps, data, org, etc.) needed to implement, update, or retire one or more capabilities. Therefore, it is crucial that a project includes the strategy artifacts needed, and does not directly jump into business and technology blueprinting. Having clarity about objectives, KPIs, business model, etc. will help in coordinating and resolving conflicts with other programs and projects who might have a less well-defined scope and program approach.

It describes the design for one transition state, and it is recommended to tailor your projects in a program accordingly by transition states related capability configurations. This allows a controlled change management going forward, while coordinating your program's schedule with the other related programs, so that the continuous operation of the implemented processes and systems for the end users can be ensured.

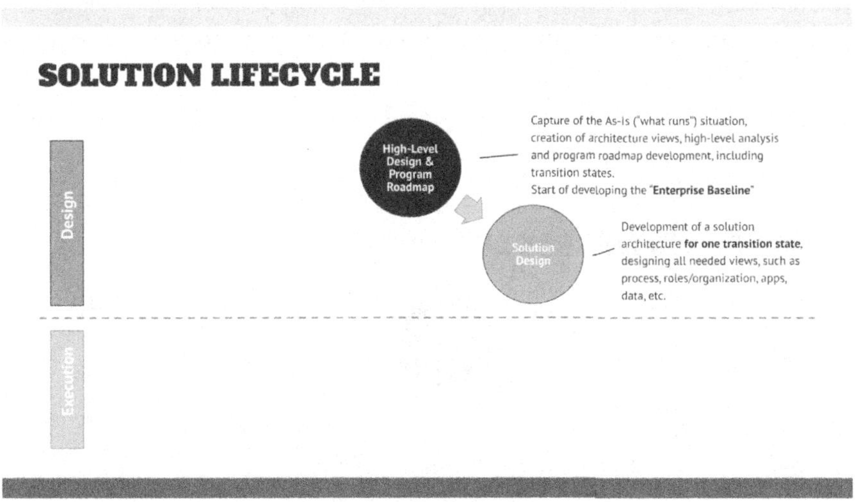

Fig. 5.3: Solution design phase

A consequence of this is that you require various roles to develop a solution – the individual domain architects (process, organization/roles, apps, data, and so on) and also a Solution Architect who is responsible for integrating the different designs into a consistent solution design. For larger programs, where multiple projects implement aspects of a capability set, you might even need a higher role – the Solution Portfolio Manager.

Once all dependencies — internally and externally — are resolved, and the design is approved, it can be moved into the next phase. In addition to this, the solution design phase can also provide input to (the update of) reference architectures, for example, when a new capability and new systems are to be implemented.

Phase 3: Implementation

This phase covers the *actual changes to the organization, systems, underlying technology, and processes.* It is mostly driven by the technical configuration of the runtime systems, but also includes the "people side" of the solution – the org change management team that puts the organizational changes in place, coordinates with work councils, etc. It also aims to enable users and admins in the new technology.

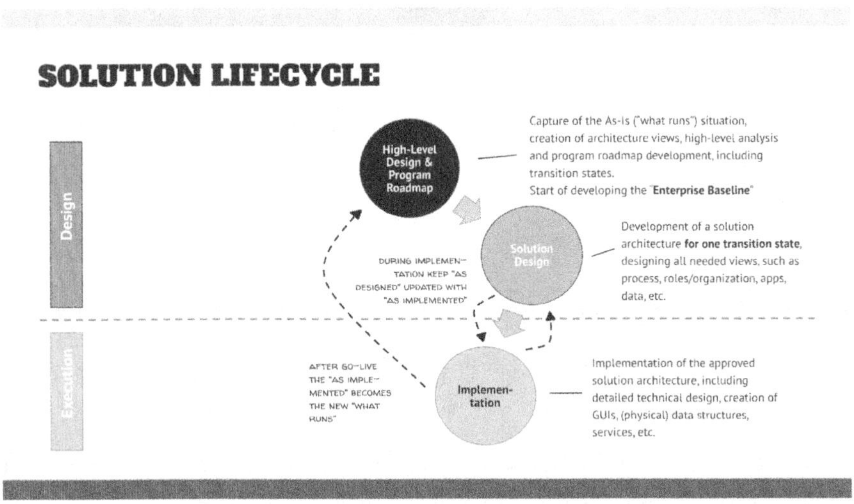

Fig. 5.4: Implementation phase

The architecture contributions in this phase are multiple:

- **Provide guidance** on how things have to be implemented, and update the solution designs accordingly if changes are needed.

- **Support the testing effort** by creating test cases for higher level tests – remember: every branch in a process is a test case and needs to be successfully executed.
- **Synchronize/upload solution designs to Learning Management Systems** such as SAP's Enable Now, so that the content there reflects the new reality regarding processes, app landscapes, and so on.
- **Update the Enterprise Baseline** with the final implemented design, so that future analysis and project definitions can be done from an up-to-date information foundation.

Phase 4: Process execution

The scope of this phase is to implement an *ongoing monitoring of the execution of the solution* on a day-by-day basis. Setting up dashboards for the different stakeholder groups — for example, operators, process owners, management, etc. — will provide real-time information measuring the service-level agreements (SLAs) for the defined processes and solutions.

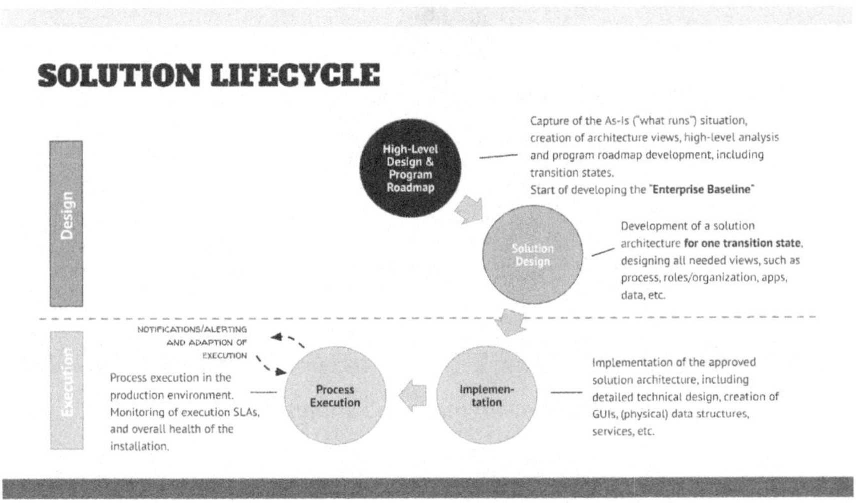

Fig. 5.5: Process execution phase

Aligning KPIs that are visualized by the dashboards with reference content (e.g., ITIL[21] or SCOR)[22] will help to compare to benchmarks and get users accustomed to process thinking.

What this phase also includes is the interaction with the execution if an SLA is not met, for example, if a delivery of a product takes too long. Notifications and alerts can be configured to trigger actions as needed.

Phase 5: Measurement and analysis

This phase is similar to the previous one, and might also use the same data sources in some cases, but the objective here is to "take a step back" and look at trends and usage patterns that show up after some time. You also might want to take a look at what your users actually do (vs. what might have been mapped out in the Solution Design phase).

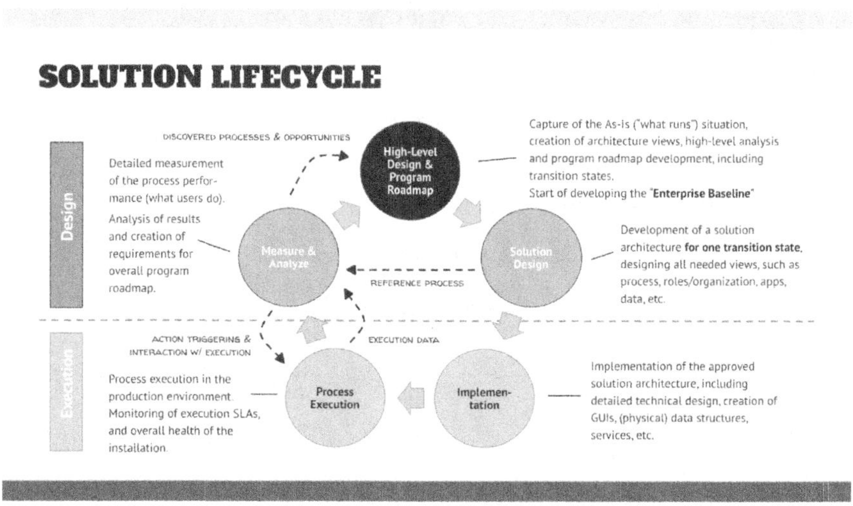

Fig. 5.6: Measurement and analysis phase

The main tools for this phase of the solution/process lifecycle are Process Mining and Task Mining (Robotics Process Discovery). Both look at usage and performance KPIs, but Process Mining typically is fed by log files from transactional systems that produce case IDs, activity names/events, and start/end time stamps. Task Mining will have agents being deployed on user machines that record what users do in which program (down to the field level) and then create process models for procedural steps – one level below the process step that you typically see in a process model (a "business-relevant, enclosed activity").

Typical Process Mining use cases are:

- **Process discovery** of non-mapped processes (that will be later included in End-to-End and functional process maps in Baseline or Projects/Solutions).
- **Process exploration** to see the variation of a process, the frequencies of activities, the throughput times of activities and connections (bottlenecks), or any other measure analysis (e.g., cost). This is the "money shot" feature of all process mining tools because it shows the "mess" that happens in reality in a spaghetti diagram. However, this feature has become a commodity and all tools do this in various degrees of detail.
- **Process compliance** – is the process executed as it is described in a reference process that is uploaded from the design repository. Good tools not only show you the compliance (are all steps done in the sequence that they are prescribed from a black-and-white perspective of the world), but also calculate the "fitness score" – to which degree is the variation of the process execution aligned to the reference process? Is it 50% or 95% aligned? How do these scores change over time?
  In addition to that, the tools will also show the "failures" in the executed processes, such as multiple start events or incorrect sequences (for example, when an invoice is received before a purchase order was created). Therefore, process compliance is also an interesting use case for (internal) auditors.
- **Root-Cause Analysis**: what are the reasons for the variations or non-compliance? Does the data show the reasons for this, e.g., resources were not provided in time, or a certain type of SKU (stock keeping unit)

or vendor creates a performance loss.

- **What/If Analysis** – some process mining tools allow changing the variables in a process to create a scenario and then have the tool run through the data to see if the outcome would be a more desirable one. This can be a valuable input for the next iteration of the lifecycle, where a future state design can be simulated in much more detail.
- **Performance visualization** – the dashboarding of KPIs that show the performance of processes, the results of them (e.g., production numbers), or the calculation of risk and compliance measures, such as the percentage of process instances in which a three-way-match has been done (the match of purchase order, goods received, and invoice). For obvious reasons, auditors are very keen on this type of information. But other stakeholders — such as process owners, application owners, as well as business line owners — will be interested in the overall performance numbers, or calculated special KPIs that are aligned to their strategic objectives.

  Typically, these visualizations are not provided as pre-packaged apps by a process mining vendor, but are built as custom apps in the process mining tool (a later article will elaborate on the roles and activities needed to implement and operate a process mining tool).
- **Actions and notifications**. Based on the analysis, workflows, and notifications can be triggered that make users aware of patterns and findings, so that the most optimal process execution can be accomplished. This could also include triggering manual reviews or control steps if the process variation is outside defined patterns.
- **Transformation dashboards** can also be created to show the improvement over time and if the defined projects have accomplished their objectives. This brings a complete new perspective to these types of projects because in the past these have been measured on "secondary" KPIs, such as the number of trained people. The actual business benefits realization happened months after the project had ended, and the project members were reassigned to the next project.

A good Process Mining tool also allows the export of discovered processes into the design platform for the next improvement circle. A Robotics Process Discovery tool provides screenshots that can be embedded into step-by-step instructions of the process models and/or be used for training purposes.

In any case, the results of this phase are a trigger for the next iteration of the lifecycle.

How do architecture tools support the solution/process lifecycle?

A good professional architecture tool stack provides features that support all solution/process lifecycle phases:

- **Enterprise Baseline**: a central repository that allows the structuring of content as defined in the meta model; instant impact analysis by using the object-orientation in the underlying database; analysis reports for repeatable use; ad-hoc queries for quick analysis; simulation; automation of [governance] processes; dashboards for insights into governance ("Are we 5% or 85% done?").
- **Solution Design**: modeling capabilities that allow the creation of blueprints in all needed architecture views; comparison features (to show the difference to the baseline current state); if available reference content to create consistent designs; analysis features to show the improvements/differences compared to current state; versioning; if available implementation tool-specific attributes and workflows for data exchange.
- **Implementation**: synchronization of blueprints with implementation tools (ideally bidirectionally to manage scope); creation of test cases and tracking of those; analysis/reporting for org change management (such as role reports as input for the creation of a training curriculum or communication items); provision of auto-generated views, such as process narratives or step-by-step views; export of content for presentations and other org change management material.

- **Process Execution**: dashboards; "live" connectivity to source systems; actions and notifications.
- **Measure and Analyze**: various analysis types for process and task mining (see above); what/if scenarios; actions and notifications; bidirectional connection to the design tool.

Figure 5.7 below shows some of these technical capabilities mapped to the solution lifecycle.

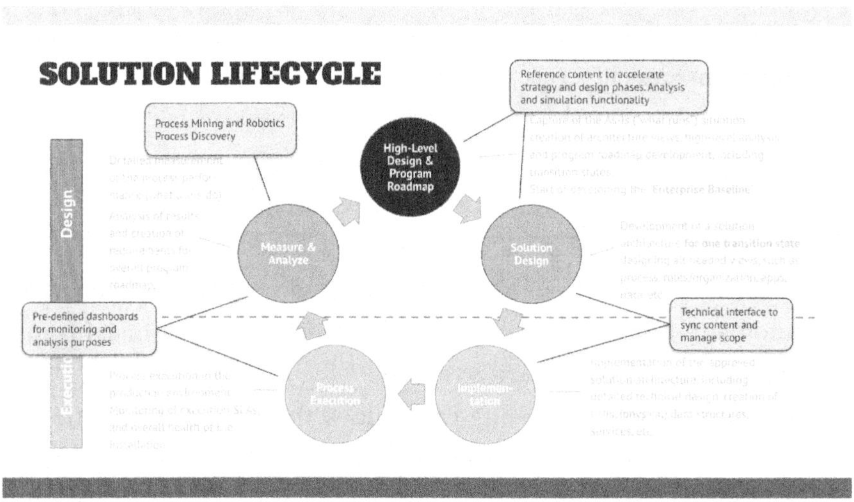

Fig. 5.7: Example tool support in the solution/process lifecycle phases

I think this chapter clarified that architectural work is not just "draw and forget" or limited to the design phase of a project. Managed well, it is a competitive differentiation by providing longer-term vision and visualization of a strategy, and the necessary information needed for solution design, implementation, and operation.

By setting up the lifecycle in the phases above, a reinforcing effect will kick in that will accelerate the maturity of design and delivery significantly.

6

The Case For Architecture Summary

In this first section of the book, I painted the "big picture" and reasoned **why** you should implement an architecture capability in your organization.

But besides the "why" there are other takeaways that you should keep in mind:

1. Business changes faster than ever. New technologies, new challenges, and social changes change the environment in which we do business today. And these "transformations" that we have to go through — either as formal projects or as the need for adaption — will not stop.
2. But there is one thing that stays a constant — your organization's architecture, which, unfortunately, is not explicitly documented in most organizations. However, if you have this (for example, a process inventory, a capability model, or an application inventory) you can analyze these and find improvement opportunities and mandatory areas of change. You then can also make the business case for those and bring the urgency for change to the forefront of your decision-makers and stakeholders.
3. A key aspect is the role of architects — are you a strategy partner for your decision-makers, a transformation agent who defines the "how" to change, or "just" a methodology or domain expert? Each of these roles are needed, but do you know how you are seen in your

organization?

4. Based on that, we took a look at the five major capabilities that you need to stand up and measure the maturity of those over time.
5. Frameworks, methods, notations, and reference content where the main focus of architects in the past and that has changed (as we've seen in the changing roles) but they still have a purpose to accelerate your architecture implementation either structurally, methodologically, or from a content perspective.
6. Lastly, I have introduced you to the Solution/Process Lifecycle as one of the two major concepts of this book (the other one is the actual architecture implementation approach shown in the next section). The lifecycle will help you to organize the different activities that you do and help you build the necessary skills that you will need in practicing your architecture craft.

II

Getting Started

7

Getting Started

Well, technically, you could get started without large parts of the content in this section. However, we are talking "approach and tools" here and at some point you will realize that you need not only some tools, but also some specific functionality beyond a simple repository.

What's in this section?

First, we cover the approach of implementing an architecture tool. This is the key concept of this book, and you will see that the rest of the chapters will detail the various aspects and give you a real-life example of what the results could look like for you.

Then we look at the tool selection, and which features and functions it should have, and which additional support criteria should be in place before you commit to a vendor.

Lastly, we investigate how the buying process of an enterprise software works and which roles (on both sides — buyer and vendor) are involved at what point. From buying the software, through the initial implementation to the operations/growth/support phase of your program.

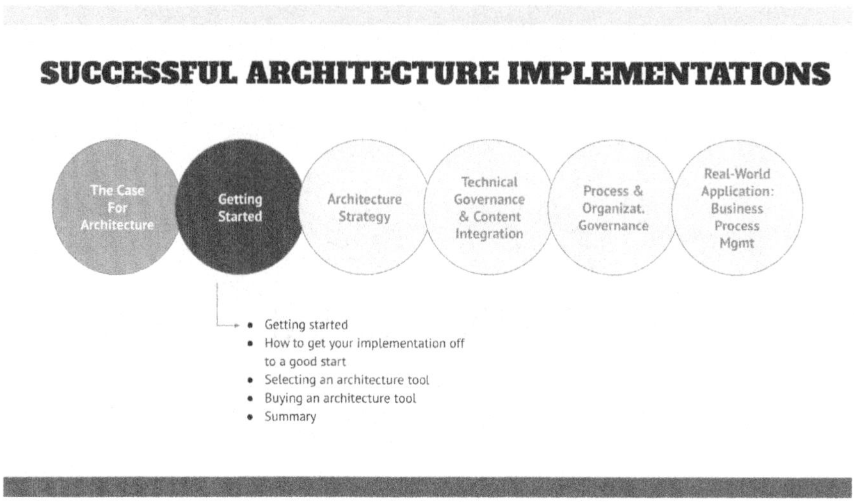

Fig. 7.1: Getting started structure

8

How To Get Your Implementation Off To A Good Start

Some clients think that buying software alone solves all their problems and that professional architecture tools are just 'Visio on steroids', which is a misconception that seriously can put the enterprise architecture implementation program into danger (for more details on the different types on architecture tools, you might want to take a look at the tool selection chapter further down below).

Enterprise Architecture Implementation

A better approach to bringing in architecture tools is to treat this program as an enterprise application implementation that is driven by the introduction of a new capability for the organization. This is obviously a larger effort that needs to be led by the business and not by IT. Doing it the right way will lay the foundation for process improvements, system implementations, business/digital transformations, and regulatory compliance.

This chapter covers the implementation of the architecture tools and is at the core of this book. A separate chapter has already covered the necessary EA capabilities that will be supported by the tools, which then can be used for maturity assessments and progression of the internal capability build.

Figure 8.1 below shows the architecture tool implementation approach in an overview:

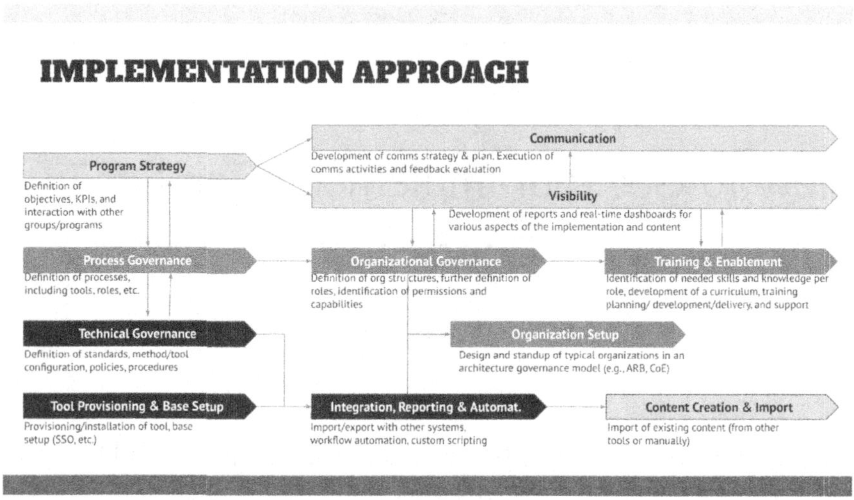

Fig. 8.1: Architecture tool implementation approach

Strategy Work Area

This work stream (light green in the graphic) aims to find the proper place of the program within other initiatives and manage it.

- The **Program Strategy** phase looks at the functional scope of the program (e.g., visualization of processes and where risk and controls will ensure regular compliance on the 'who-does-what-when-where' level, typically depicted in BPMN diagrams), and identifies the 'sources of truth' for information, as well as where the architecture tool becomes the main source of information.
  This will then also determine how the program will interact with other programs and/or business-as-usual org units that are responsible

for systems, as well as the necessary IT/development support (and standards) that are needed to create integrations. This is shown in blue at the bottom of the graphic. Obviously, this will be captured in a "living" Program Strategy document that can be used as a reference in the future of the program.

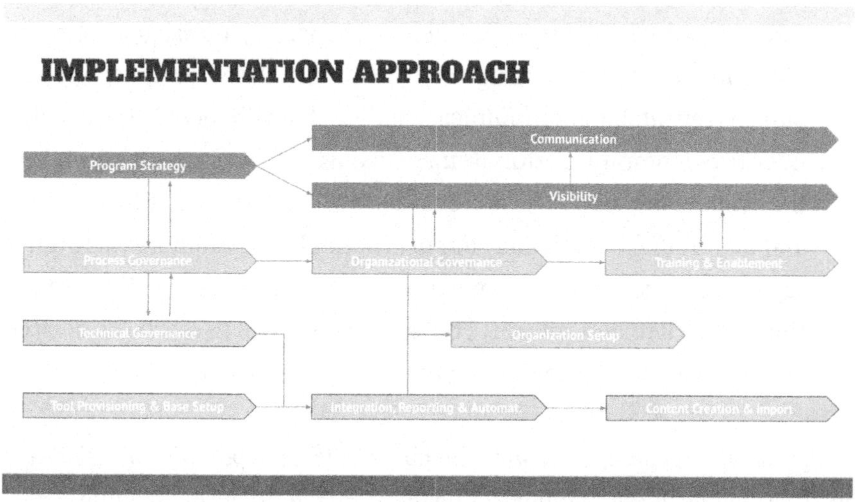

Fig. 8.2: Strategy work area

- **Visibility** — The next step then is to define and formally describe the objectives of the program over multiple releases, as well as the KPIs to measure success, and identify where they come from. This, and the identification of stakeholder information needs, sets the foundation for the development of reports and real-time dashboards in the Visibility phase that support the management of the program.
  All artifacts will be done in a Solution Architecture database, which will be extended in other work packages, and serves as the 'meta' information source that will be used when interacting with other

entities.

- **Communication** – As we have seen in the "Which Capabilities Do I Need to Build My EA Practice?" chapter above, you need to consider the three areas of successful implementations: Content, Governance, and Adoption. One of the big things in the Adoption area is to start communicating as early as your program was approved.
  To do this successfully, you might want to start laying out the foundations for it while developing the strategy in the form of a Communication Strategy, which covers stakeholder groups, key messages, communication and feedback channels, and so on. Based on that, you can build a (repeatable) Communication Plan that will include the specific measures/communication items, such as a newsletter or event, and the responsible people to execute on it. The last step is to establish and maintain feedback and support channels, for example via forums, regular feedback collection meetings, and so on. The objective is to use the feedback to determine if you are still on track to reach the desired business capabilities laid out in the Program Strategy.

At the end of this phase, the foundation for stakeholder reporting and visibility is in place and the initial communication has started to create momentum.

Governance Work Area

A second 'housekeeping' item is to define the governance that applies to the program (dark green in the graphic). It typically is set in four work packages that interact with other work packages, and all designs are developed in the Solution Architecture database.

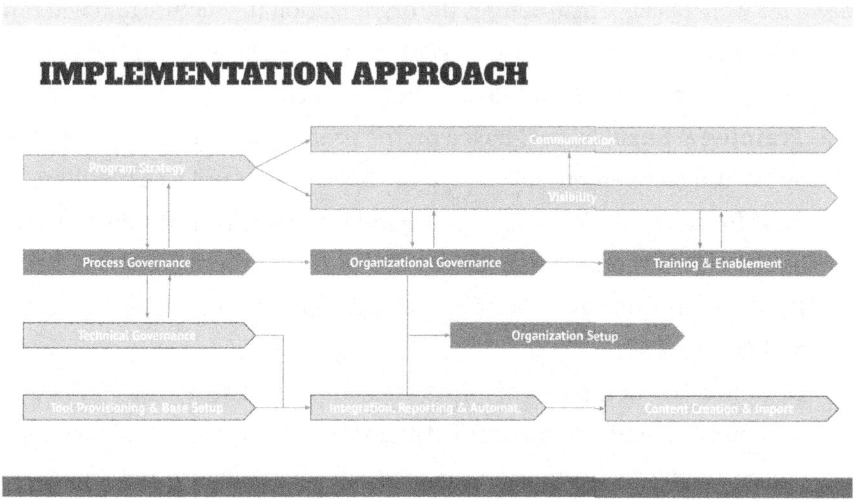

Fig. 8.3: Governance work area

- **Process Governance**: The definition and documentation of the processes (including "auxiliary information", such as to-be-created artifact, used applications, etc.) that users, workflow participants, admins, and stakeholders do. These cover processes such as the creation or import of models, reviews and approvals, or the management and operations of technical interfaces that will be created with other systems. In addition to these policies and compliance requirements, such as the level of detail, or the frequency of architecture artifact reviews/certifica tions will be developed in this step.
  Typically, the models created in this phase consist of one high-level Value Chain Diagram with assigned BPMN processes underneath, which will include the business steps, roles, applications, etc. – the 'who does what when where' level processes of the program.
- **Organizational Governance**: Based on the defined processes, roles are identified, and permissions and privileges will be defined and administered in the tools. The main artifact for this is a license/per mission/privilege matrix that includes every role.

This work package also defines the organizational structure that will run the tools in production, such as a Center of Excellence (COE), a standards group, or an Architecture Review Board (ARB).

- **Training & Enablement**: Based on the process and role definition, the skills and enablement needs will be determined, and a curriculum be built for each role. This not only includes the formal training, but also additional support measures.
  Based on this design, the training, and support material will be built and delivered.
- **Organization Setup**: In this work package the identified org units will be stood up, candidates identified and selected, and the initial meetings supported by the program management until the org units are self-sufficient in their operations and planning/management.

At the end of this phase, the "rules of engagement" are defined, the needed org units/teams stood up, and users are trained in the use and administration of all systems used in the program.

Technology Work Area

This work stream (dark blue in the graphic) is typically the only one that clients think about when it comes to the implementation of an architecture tool.

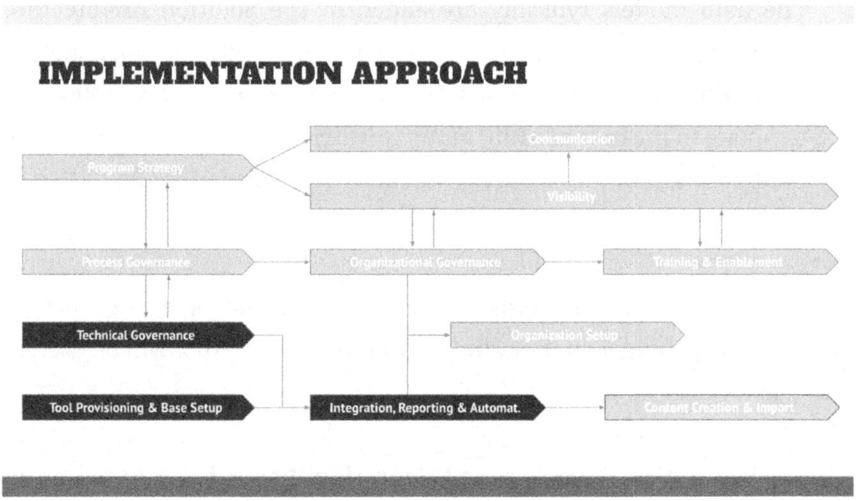

Fig. 8.4: Technology work area

It is typically divided into three work packages:

- **Tool Provisioning & Base Setup** aims for making the tools available to the program. Depending on the hosting (cloud or on-premises) this will include different steps and participants, but typical activities include the setup of single-sign-on, connections to the network, integration with email servers for notifications, etc. At the end of this activity, the tool vendor hands over the keys to the tool, and the program takes over responsibility for the tool. While the vendor and/or the IT organization manages the technical aspects of the tool, such as backups, user group creation, etc.
- **Technical Governance** determines the configuration of the tool and the underlying method decisions. This includes the creation of data models and interface matrices (to other systems), as well as a meta model that shows the levels and views, as well as the chosen artifacts on their intersections. These artifacts (typically models) are defined by their notations, objects, attributes, etc. This work package also includes the database setup and structures within each database.

The data models typically are added to the Solution Architecture database, while the meta model is created in a Configuration database. This allows the creation of the tool configuration files (filter and template in the case of ARIS) automatically by running a wizard on the Configuration DB, which saves the time of the manual configuration of each item.

It also facilitates the workshops for the data model and meta model design and gives stakeholders a first glimpse of how the new tool will be used in other solutions.

- The last sub-work package of **Integration, Reporting & Automation** picks up the deliverables of the Strategy and Governance phases and includes the creation of scripts for technical integration purposes, visibility reports, and the automation of (governance) workflows within the architecture tool stack.

While doing this, leading IT practices, such as a defined development approach, or the use of multiple environments for development, test, and production should be considered. Stay away from implementation partners who see this development as 'one-offs' that do not have to follow these practices.

At the end of this phase, all tools in the program are stood up, configured and ready to use.

Content Work Area

The last step in the implementation (light blue in the graphic) is the creation and import of existing content and/or reference content from standards organizations, such as APQC[23] into the Content (Work) database.[24]

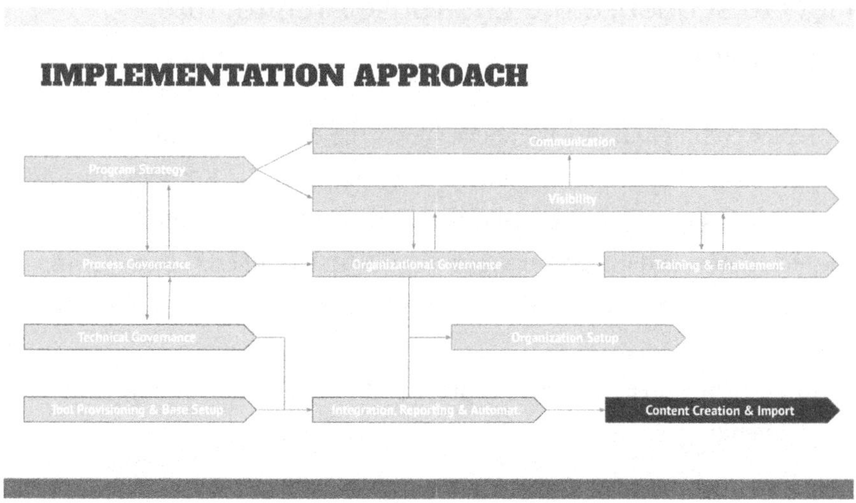

Fig. 8.5: Content creation and import work area

Existing content might exist in various formats—Excel, Visio, *.bpmn files, etc.—and need to be imported into the tool. This can be a larger effort, since the main work is not initial technical import (this is typically done well by the tools), but the normalization and integration of the content into existing hierarchies—like the other topics above, this will be discussed in another article later.

Another part of this work stream is the import of reference data from upstream systems, using the interface scripts from the Technology work package. This data typically comes into the tool as catalogs, and they might have to be rationalized. Other imported content, such as processes, need to be reviewed to see if they used the correct content from the system of truth. This is in case if there are discrepancies or needs for consolidation of objects to ensure consistency of the database and the content structure.

At the end of this phase, the architecture tool is ready for a roll-out to a wider audience.

How to sequence the Enterprise Architecture implementation

The color coding in figure 8.5 above shows the type of roles that do the four work areas. Program management is working on the Strategy, business consultants support the governance design and technical folks take care of the technology and integration/reporting. A first set of end users will import existing content and create initial structures within the Content/Work database.

However, not all activities take place at the same time. My suggestion for the sequencing is:

1. Program strategy, and tool provisioning & setup (assuming the tool selection is already done).
2. Definition of process and organizational governance, as well as technical governance; start communicating.
3. Build of integration, reporting & automation and visibility.
4. (In parallel to 3) Stand-up of the org units, and training design and delivery to the initial first end users, who will become power users when the tools roll out with a wider audience.
5. Import of existing and reference content.

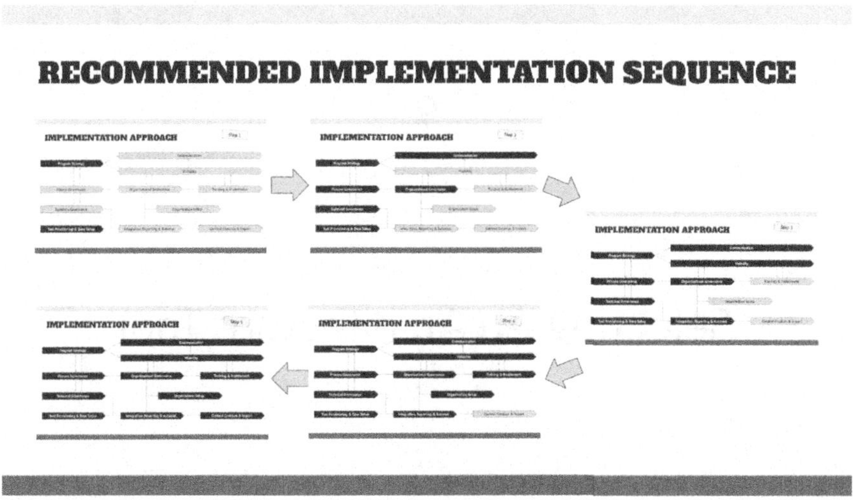

Fig. 8.6: Proposed sequence of an Enterprise Architecture implementation

Conclusion

Implementing an architecture tool is more than just subscribing to a tool and assigning licenses to users. It needs a plan that is aligned with the organization's strategy and aims to build a capability instead of just creating artifacts that are not embedded in the process of delivering solutions. This may be simple process changes (BTW, they are never simple when you think of the adoption challenges or showing regulatory compliance), change of an organization, or implementation of multi-million dollar technology.

You should treat the implementation of your architecture tool therefore in the same way and allow yourself enough time to keep the three pillars of an architecture tool implementation in balance: Content, Governance, and Adoption.

9

Selecting An Architecture Tool

Setting up an architecture/process management capability is a large endeavor, and the choice of the enterprise architecture tool (or the refusal/delay of a tool selection) can determine the overall success of the program. If the tool does not provide the necessary functionality, the value can be diminished or not realized, and additional purchases might be necessary to compensate. This brings the overall cost over the cost of a more complete – but pricier in the beginning – alternative. Or the tool chosen is just for one user group (the nerdy analysts and architects) and the implementation fails because the tool is not accepted by the rest of the organization.

Categories of Enterprise Architecture tools

Architecture tools that provide varying functionalities and capabilities, and you can buy more modeling-oriented tools, or more database-type tools where information is captured on forms and not in diagrams. Modeling tools are more flexible than DB-driven tools if they fall in the latter category, shown below:

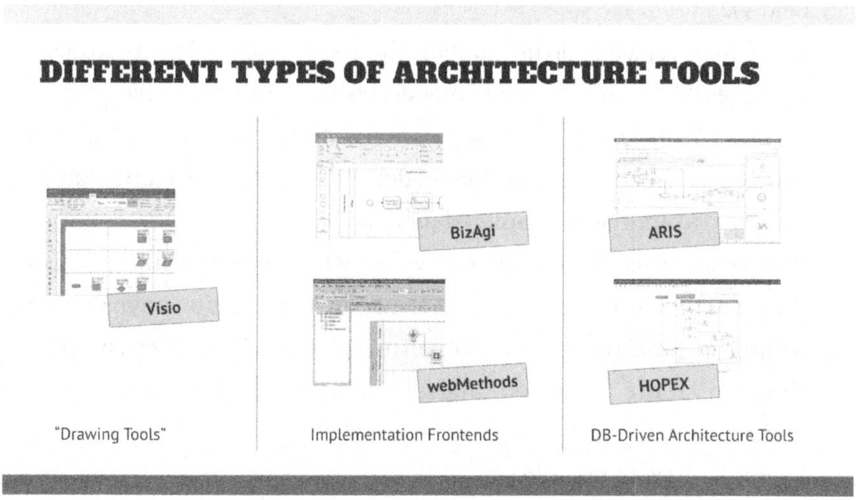

Fig. 9.1: Types of architecture tools

- **"Drawing Tools"** – Allow users to "draw" a diagram, but typically create standalone artifacts that are not connected, and therefore do not allow model analysis or the creation of a model structure that adheres to a framework such as The Open Group's Architecture Framework (TOGAF).[25] Further, these tools do not provide the functionality for users to develop a process model hierarchy that allows users to drill-down into lower layers of a model or connect models for essential navigation. In general, drawing tools do not provide support for semantics that help a user to model in accordance with a specification (such as BPMN 2.0), and do not provide automated semantic checks to highlight non-adherence to desired modeling specifications. They also allow for a lot of "artistic creativity", which is something that you might not want. Example: Microsoft Visio[26]
- **Implementation Front Ends** – Allow the visual modeling of processes, while requiring the configuration of an underlying execution engine. As such, not all artifacts that are typically needed for a complete process design are available (see the "Elements of a process" graphic/figure 9.2

below).

These tools typically do not support the creation of model structures/ hierarchies or the connection of models because the executable processes that they help configure are self-contained with well- defined start and end events. In other words, the models "pile up" on each other and are managed as individual, isolated models.

Process analysis in these tools is typically restricted to single processes that are currently executed (performed in a production environment), using the runtime data available from the production environment. However, more advanced analysis methods, such as 6 Sigma, are typically not supported by this class of tools.

Example: IBM Blueworks[27]

- **Database-Driven Enterprise Architecture Tools** – These tools are built for the specific use case of designing and analyzing any desired type of process and architecture views. As such, these tools possess the capability of managing numerous models in a multi-user environment. These tools often maintain tens of thousands of models in a database with numerous users accessing the database simultaneously, which allows them to scale well throughout an organization.

  From a functional perspective, these tools support standard industry frameworks and notations out-of-the-box, which enables efficient implementation while also providing the modeling consistency of a standard. This is supported by automated syntax checks, and other automation possibilities (e.g., establishing governance procedures with automated approvals processes).

  For large, complex organizations interested in establishing a BPM/EA capability, this class of tool provides significant advantages above others.

  Example: ARIS,[28] HOPEX[29]

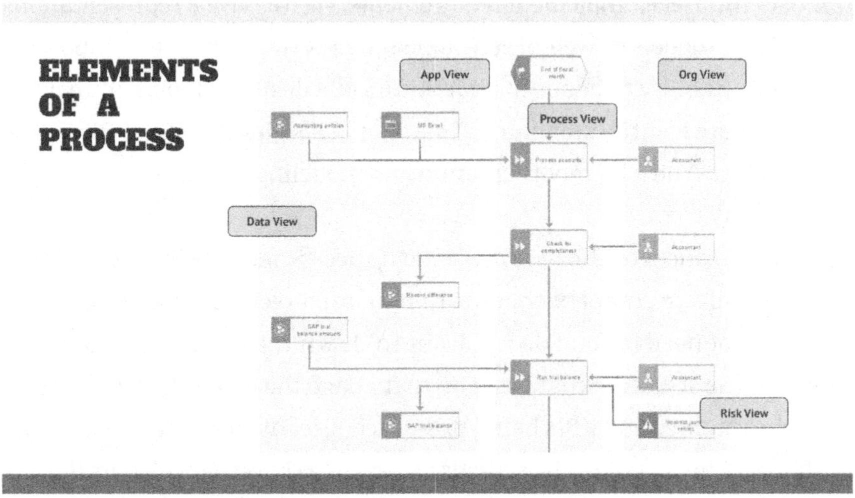

Fig.9.2: Different elements of a process to be captured in an architecture tool

General Enterprise Architecture tool capabilities

When evaluating a process/architecture tool, the following criteria, separated into the four categories of user focus, functional requirements, methodology, and technology, should be considered:

User focus

The enterprise architecture tool should be able to support various end user types/personas, and provide the information for them in the right format, so that not every user needs to be a "full modeler" to use it. This means that, for example, management stakeholders will interact with dashboards and tabular views, while analysts mostly will create and modify diagrams. Regular end users might be only interested in narratives and step-by-step instructions that support the work at hand and will be displayed on kiosks or tablets.

- **User Interface**: implement requirements, such as visual representations of large models or web-based publishing (portals), as well as modeling and the ability to adapt to an institution's desired "look-and-feel" or corporate identity standards. This should also include What-You-See-Is-What-You-Get report creation and modeling support.

Dashboards and workflows: One critical success factor will be the ability to communicate complex content to non-architect stakeholders. These users do not need to (and do not want to) learn the details of a modeling tool, and the structure that is set up to manage the content. They want to have the findings and KPIs being presented directly in easy to understand dashboards, or be guided by screens as part of a (governance) workflow.

- **Reporting**: Provide robust reporting and dashboarding capabilities, allowing the analysis of the BPM/architecture content in real-time or on demand. This capability is especially necessary for non- technical business users who require readily accessible overviews of results without interacting with the underlying database.

Functionality

A good tool should support the full lifecycle of the content (see the "Solution Lifecycle" graphic/figure 9.3 below), and either provide the necessary functionality directly in the tool stack, or have pre-built interfaces to specialty tools that provide the desired functionality.

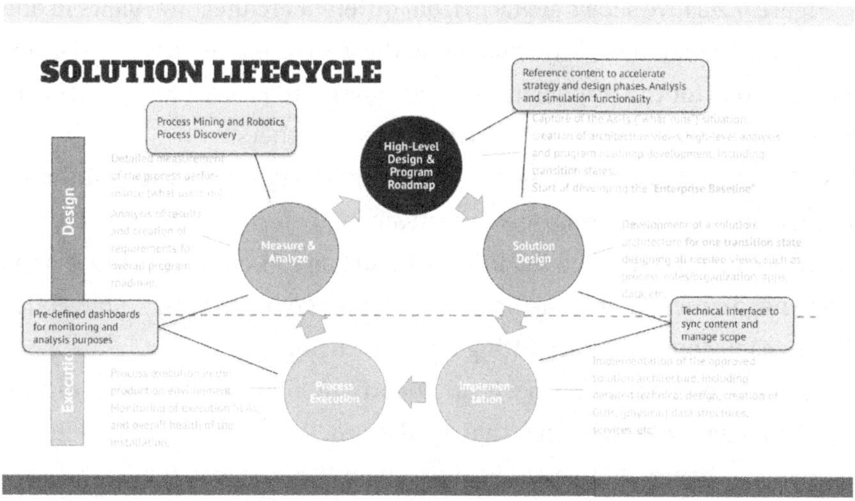

Fig. 9.3: Example tool functionality support in the solution/process lifecycle phases

The artifacts of the specialty tools should be integrated seamlessly in the architecture tool.

Methodology

As shown above, the preferred type of tool – a database-driven tool — should have a built-in methodology/meta model. A meta model determines the available model types, the object types within the available models, and attribute types and connection types.

Technical requirements

The selected tool will be accepted better as a System of Record if it can be integrated into an existing IT landscape and is managed like a "real IT system" (vs. a drawing tool). This will also bring the architecture/process management team to the table, and position them as a main player in the organization.

Figure 9.2 above shows artifacts from different architectural views in one diagram. A good tool allows managing these in different views (and grant permissions for the responsible user groups), while also allowing to bring in artifacts that are read-only and managed by others.

Only some database-driven tools bring these key capabilities, and some of them have included third-party software in their core product, or created interfaces to the market-leading niche players in an area. For obvious reasons, an inclusion is the preferred way to go because the tool vendor has already done the heavy lifting for their customers.

The main purpose of the tool is to support the enterprise architecture/process management capabilities that are desired. We already discussed this in a previous chapter, but here it is to remind us what to look for.

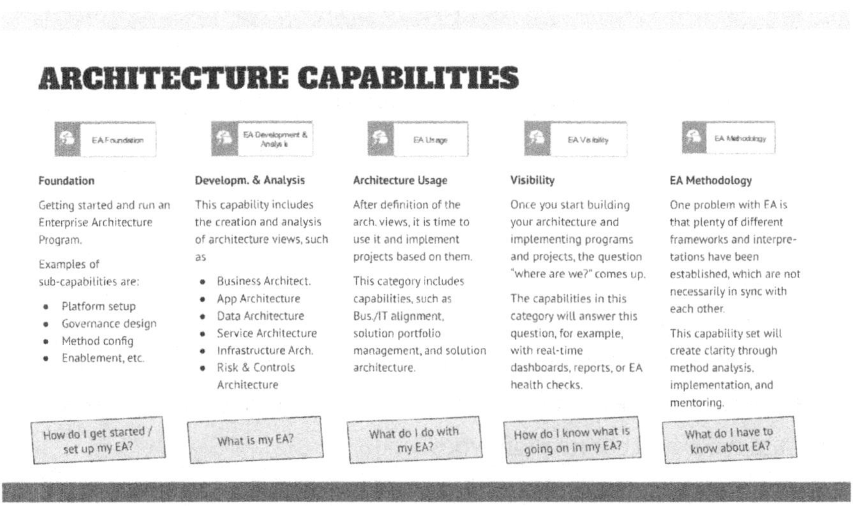

Fig.9.4: Main Enterprise Architecture capabilities

Technical criteria for architecture tools

Going forward, we will take a look at more advanced/technical criteria that can make your tool a success, or a "one-trick pony" that will end on the shelf soon. Users will not see a positive value compared with the effort that they put into adding content into the repository.

Extensibility

When choosing a modeling tool, the expectation is that it supports a wealth of industry-standards, but also allows users to enhance the capabilities to accommodate, capture, and analyze institution-specific information.

These enhancements are enabled by two features:

- An **extensible and flexible meta model**, which allows users to create new model types, object types, and attributes to enable specific use cases.
- A s**cripting/reporting capability** to perform analysis, support modeling and administrative tasks, and create interfaces to other tools (covered in the "Interfaces" section below).

A meta model provides the structure governing information connectivity within the tool, and defines the relationships between the models, objects, and attributes that are available.

When considering a tool, consider the ability to enhance the meta model with client/project-specific additions, as well as the ability to adapt/customize the meta model for individual user roles (e.g., business analysts creating process models and org charts, an information architect who has to create logical and physical data models, an application architect managing a system portfolio, etc.). Ideally, the selected tool will provide all of the above functionality.

Figure 9.5 below shows the comparison between three tools in these two areas as part of a client project:

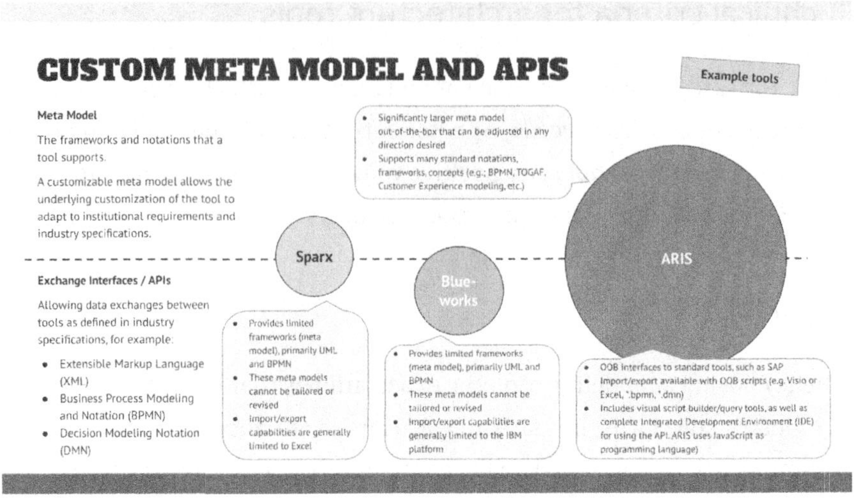

Fig. 9.5: Technical capability comparison between three tools

As part of a Proof-of-Concept we extended the ARIS meta model to accommodate the client's business requirements and upstream system specifications, by creating new attributes that allowed a 1:1 mapping of data fields in the interface, while still adhering to a client's chosen standard notations (e.g., BPMN).

Additionally, ARIS was used as the design tool for the interface development, interface governance processes, application landscapes, logical and physical data model, and security/permission design. Other Process management tools such as Visio or Blueworks could not have provided the above functionality, as they do not allow users to edit and enhance the meta model as ARIS does.

Interfaces and connectivity

In addition to allowing the extension of the meta model, architecture tools should be evaluated on the ability to readily connect to other tools and applications.

Common considerations regarding connectivity include:

- **Out-of-the box integrations** with other tools (such as SAP or ServiceNow)
- Ability to **import/export common formats** (Microsoft Excel, Adobe PDF, CSV, etc.)
- An **open Application Programming Interface (API)** that can be used to provide information for use in other applications
- An **extensible meta model** (see above)
- A **scripting engine/development environment** that allows to create the necessary scripts
- Support of **standard programming languages and industry exchange formats** to avoid vendor dependency issues and to leverage existing organizational knowledge and skills

In the project example above, we compared the out-of-the-box integrations available for a customer – Sparx,[30] Blueworks Live, and ARIS, and they showed significant differences. Sparx and Blueworks have limited integration ability compared to ARIS. For example, Blueworks can be integrated with other IBM products. ARIS, however, has the following integration abilities:

- **Tools developed by Software AG**, such as web Methods (a Business Process Automation Software),[31] Alfabet (an IT Portfolio tool),[32] or Aware (a dashboarding tool).
- **Enterprise Resource Planning systems**, such as SAP and Oracle ERP, or Configuration Management Database Systems (CMDB, such as ServiceNow).
- **Import and export capabilities from various tools and formats**, such as Excel and Visio (mostly used by business users) or *.bpmn, *.dmn, and .xml file formats for exchange with existing IT systems.

ARIS also has an open API that allows the use of mobile devices to be used with ARIS, using standard web technology (HTTP, REST, JSON, etc.). This API is not meant to replace a more traditional, stable integration, but enables working with the process information everywhere based on geo-information or QR codes. Use cases for this functionality might be the context-specific display of process descriptions based on a region or product information in a warehouse or retail outlet. These instructions then can also include multimedia items, such as videos or pictures, for additional information. The mobile API also includes governance activities, such as reviews and approval workflows being run on a mobile device, or the capturing of a potential risk at the point of execution during audits.

When creating the interfaces, a tool shall bring either its development environment or integrate seamlessly with standard IDEs (Integrated Development Environments). Ideally, the IDE is an integral part of the tool, and no additional software purchase is needed – again, let the vendor do the heavy lifting in supporting the full development process.

To integrate or exchange information with other tools, the tools require scripts and reports. In the project example above, ARIS showed multiple user interfaces capable of creating these scripts/reports:

- A **visual query editor** that allows users to create queries by simply grouping items together, step-by- step in the graphical user interface: the result can be shared with other users, and creates a spreadsheet model that can be exported into Excel as needed.
- A **WYSIWYG (What you see is what you get) report editor**: While more complex than the query editor above, the WYSIWYG report editor allows advanced business users ("power users") to create reports in a visual environment without having to write code. Reports can import/export data, automate tasks within the tool, and can be tailored to update existing content or create new content in the database without human interaction.
- For more complex integration and automation scripting needs, an **integrated development environment** that uses standard programming languages: using an open language (versus a proprietary programming language) increases the likelihood that an organization may have employees with the requisite skill set to work with the tool, reducing

the need for extensive retraining and/or vendor dependency.

Queries and reports can be used as data sources for the included dashboarding component and can be refreshed upon access to a dashboard to facilitate real-time data updates.

Deployment options

Most tools can be deployed in a client data center or in a cloud-hosted environment. If cloud hosting is not desired (for security or other reasons), a common election is for an organization to deploy the tool within its data center. This requires a server and table space, as well as open network ports, so that the clients can communicate with the server. As an example, for ARIS there are two client deployment options:

- **Download clients** or **locally installed clients**
- Modeling **directly in the browser** (the direction in which the tool will go)

Even though option 2 is the desired one due to the benefits of deploying the tool to the users (there is nothing to deploy on the users' machines), often there is a need for a hybrid approach, which offers both options mentioned above. For example, selected users may use a locally installed client that includes administrative capabilities (user management, database backup, etc.), or advanced functionality (e.g., simulation), while business users will achieve adequate functionality utilizing a web browser modeling environment.

In addition to this, some tools' servers can also be installed stand-alone on a single PC. However, this is not recommended as a deployment option for a multi-user environment, but rather considered a temporary solution in case a user has to make changes while traveling without access to a central server.

A cloud-hosted version of the tool typically offers the same functionality as the on-premise version, and depending on the package chosen, there may be limitations regarding interfacing with other systems (private cloud vs. public cloud hosting). Accessing the cloud server by a user is identical to an on-premise installation.

One example is IBM's Blueworks which is a public cloud-hosted environment that may have restrictions regarding interfacing and access as mentioned above.

Technical Support

When selecting a vendor, you might want to look if they offer support for both deployment options detailed above, have a technical support program, and offer professional services for the company's products. The technical support is typically included either in an annual maintenance fee for on-premise installations, or in the cloud subscription fee.

In addition to this, there should be a very active user community. For example, Software AG's ariscommunity.com has more than 750,000 members as of 2024. They also host annual user conferences and organize local user group meetings, where users can exchange experiences and remain informed of new features and developments.

In addition to the tool vendor support offerings above, other third-party companies/tool vendor partners maintain broad expertise and offer services around Enterprise Architecture and Business Process Management end-to-end. The size of this ecosystem might influence the selection decision.

Analyst recognition

The market for architecture/process management tools is mature, and covers the process improvement, automation/integration, and analysis use cases. This is recognized by the major analysts (Gartner and Forrester), reflected by consistent ratings of vendors in the "leaders" analysis section.

Until 2013, Gartner had a specific "Business Process Analysis" (BPA) Magic

Quadrant. Since 2013, the structure of the analysis reports has changed, and the BPA analysis was included in two areas:

- **Business Process Management Systems**, that include not only a modeling and analysis environment, but also the runtime systems[33]
- **Enterprise Architecture tools**, that include IT application portfolio management in addition to modeling capabilities

This new structure does not necessarily match a client's current evaluation need for a pure process management tool, but some "old guard" tools from pre-2013, such as ARIS or HOPEX platform, are still around and in active development. They are either integrated or connect to the tools in the leaders area of these new quadrants. One example that stands out here is HOPEX which covers both scenarios in one repository. It allows a more model-oriented approach for business users (e.g., process models), while also providing an object-oriented approach for IT users (e.g., management of individual applications or pieces of technology) in one repository, while other competitors require a synchronization/interface between two tools.

Closing summary

When looking at selecting an architecture/process management tool, it is a good idea to not only to look for the functionality required in the current moment. Looking outside of this and evaluating technical integration capabilities, support offerings — either by the vendor, user groups, or partners — are important criteria as well, since the success of the tool regarding the acceptance by the users is critical.

Buying a professional architecture tool because it is "Visio on steroids" is never a good idea, and providing a tool will not guarantee the implementation success ... but this is a topic for another article.

Bonus: To make things easier for you (and as a thank you for making it through this semi-boring chapter), I have created a Google spreadsheet[34] that you can use for your tool evaluation efforts. It is read-only, so save a copy of it first. ;-)

10

Buying An Architecture Tool

My assumption is that most readers of this book are working in a "regular" organization and are looking for a tool that will help them to do their work. They are not procurement specialists, or work for a vendor.

This chapter will look at the process of buying an enterprise tool and also look at the "other side" of that process — the vendor's perspective. This might give you insight into why they react how they react to your questions and demands.

Who is involved in the buying process?

There are different players involved in this process — on the buyer side and on the vendor side. Let's take a look at them:

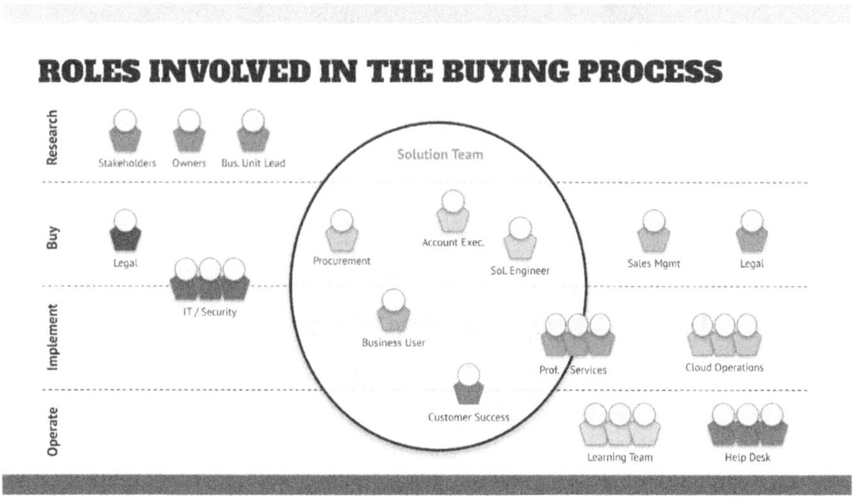

Fig. 10.1: Roles involved in the buying (and implementation) process

- The first one is the **business user**. This can be -theoretically- anyone in the organization who has a problem to solve. Most likely, you will be part of a BPM/EA group or work in a project that needs to get a solution designed or improvements identified.
- If your organization is a bit more mature, you already might have some **owners** defined who are either responsible for an area (like process owners, or application owners). Or you have set up whatever you could set up in an enterprise-wide organization, and you run out of "juice" with your office files and the management of the artifacts becomes too much.
- I also have seen **business unit leaders** who jumped on the process improvement / optimization bandwagon, and are looking for a tool that they can "grow into". Occasionally, these folks are not hands-on involved in the buying process, but need to get educated so that they can approve the expense and provide their sponsorship and support to the project.
- **Purchasing people**. The pros in your organization to get money spent

and negotiate the best deal for products and services. They can be a pain at times and be "overly formal", but be assured that they know how the game is played, and you want them on your side.
- Your **IT people**, who already might have people who call themselves architect or who need to approve the inclusion of a new software in the company's ecosystem. You need these folks because you need their operations support and resources to have your tool run reliably and performant in case of an on-premise installation.
  This group also includes the people in your organization who manage (IT) security.
- **Other stakeholders** like Risk and Compliance people who might have specific requirements for the software that must be included.

On the vendor side, you typically see these roles:

- The **Account Executive**. This is the salesperson who is your go-to contact during the buying process. These people manage the initial scoping conversations — and the better ones will give you initial demos — and will assemble the vendor team for your request.
- **Solution Engineers**. These are the rock stars in the process, and you want to build a good relationship with them. They listen to your requirements, translate that into solutions built on the software, and are also the knowledgeable tech people who will educate you and your colleagues on the abilities of the tools and the concepts behind them. Solution Engineers also build the demo and proof systems and will accompany you during proof of concepts, etc.
  And the better ones will also tell you when their software is not the right thing to solve your problem.
- **Sales Management** are the bosses of the previous two roles, and they get engaged when it comes to non-standard sales situations. Maybe because you are a large organization with special requirements, or you dangle a massive carrot in front of the salesman (remember: most of them are "coin operated"). These roles on the Director or VP level will

also be the ones who will make the case when it comes to discounts, etc. Sometimes larger organizations have this bundled in a (global) deal desk ... the counterpart to the procurement organization in the buying organization.

- **Legal**. In both organizations, there will be lawyers who want to give their stamp of approval when it comes to contracts that are not standard. These folks are also involved in the beginning when it comes to non-disclosure agreements (NDA) so that you can even start sharing information.
- **Professional Services**, or a **partner** organization, who will provide the offer for the implementation and other services that you might request.
- **Customer Success Managers** (CSM). If your deal is over a certain size or non-standard, you might get a CSM assigned. They will take over as your day-to-day contact after the sale is done, and you are either in the implementation or, thereafter, in the production phase. CSMs will work with you on success plans and share their knowledge from other implementations, and in many organizations, are also responsible for renewals.
- **Learning organizations** — the better vendors have dedicated groups for learning and enablement. These typically offer a wide range of education products and services, that can span from self-paced e-learning, through instructor-led training, to formal certifications. They also might maintain a knowledge base for customers, but typically, you find the learning specialists in these organizations, and they might be able to answer "standard questions". If you have specific questions, the professional service people and the Solution Engineers might be the better contacts. After the sale, the CSM will be your go-to person.
- **Cloud Operations** is the group that will speak with your IT /Security folks and provision the software and do some basic configuration like the single-sign-on setup.
- **Help Desk**. Typically, vendor organizations have a three-tier help desk setup: a first level might have dedicated individuals who are available at certain times and take on your questions and might be able to help

you with them. Depending on your vendor's size, this can go from "business hours Mon-Fri" to "24/7/365" with help desk people in different locations and timezones.

If the first level can't answer your questions directly (the good firms have built knowledge bases that they, or even you, can access), your request will be forwarded to dedicated second level people, who are specialists in the different areas of the software. If the problem goes deeper, the third level typically is the product management/developmen t group, and you might have found a bug that needs either an immediate fix (in case of security issues) or go on the planning board/roadmap for future releases.

In either case, the first level help desk will be the main communication partner for you and manage your requests via tickets (and in some organizations you can look up the status of your ticket by yourself).

The buying process (and thereafter)

Buying an architecture tool is typically an enterprise software process that we will discuss below. There might be some vendors who offer self-service without engagement of humans, and that might be enough for an individual project. Typically, you will not get support past the learning offerings and are on your own to learn via the documentation.

The process for buying an enterprise-level architecture tool is shown in figure 10.2 below:

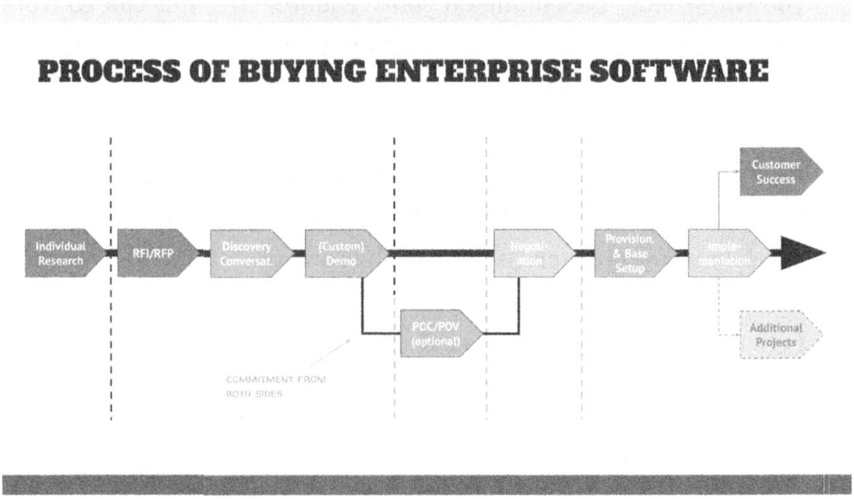

Fig. 10.2: Process of buying enterprise software

Internal Research

The process starts with the business user doing his or her research and the self-education on use cases, features, and other specifications, such as tiered offerings etc.

This phase typically is very unstructured and people come up with a list of things (like: "must model in BPMN", "must support framework XYZ", "must integrate with existing system ABC") and an initial market list. However, only a few organizations that I have worked with have gone the extra step and already thought about the different personas/roles that we will speak about in a different chapter below. They have then assembled what life with the new tool will look like for each persona.

You also want to include the other stakeholders from above in this early research so that you learn about their needs and requirements; especially regarding security, cost, or budget.

One thing that Google has researched was that it takes 11 interactions with a brand and 7 hours of learning about the tool and the vendor before customers even attempt to reach out to a vendor. So, vendors, you better have good content on your website and make material available in the form of podcasts, webinar replays, marketing material, implementation guides, etc.

As indicated by the vertical red dotted line, this might be the first point in the process where a potential buyer checks out of the process and does not include your tool on the list for the next steps. And a vendor might never learn about this opportunity.

RFI and RFP

These acronyms stand for "Request for Information" and "Request for Proposal" and basically mean the same except that an RFP would like to see specific estimates for all aspects of an implementation (in dollars), while the RFI might only include high-level "ballpark" numbers. The RFI can be issued at any point, while the RFP typically comes a bit later, when the business users have made a pre-selection of vendors and the requirements are more solid.

Typically, these docs will be issued by a procurement organization and there has been an internal process of purchase request been done in the buying organization, including approvals by leadership. In the majority of cases, a buyer organization sends over instructions and a big Excel file with the requirements listed by functional, non-functional, security, etc. requirements. There also might be a form to be filled out that will hold the pricing information. The expectation is that a vendor will fill out these documents, but in almost all cases I advise against that.

Discovery conversation(s)

The better way is to have a conversation with the sales people ("discovery conversation") in which you give the vendor a wider overview of what you want to accomplish. This way, the Account Executive can assess if their tool is a fit on a higher level, and which team members need to be assembled.

Then a second call (and maybe additional calls) that includes a Solution Engineer should be scheduled. The objective is that this is not just a checkbox exercise. You have open conversations and define a solution that not only includes the technology, but also takes all other aspects of an implementation — which we will discuss further below — into consideration.

In my experience, this phase needs deeper conversations and words will be interpreted differently. The ideal outcome is that there is a shared understanding (ideally in writing, which includes diagrams) of the solution that will be implemented. This includes everything from the objectives, through the technology, implementation, and measurements on a high-level.

(Custom) Demonstration

This is a phase that might run in parallel to the discovery and solution shaping conversations.

It begins with a standard demonstration. Most vendors do have their use cases implemented in their tools using a dummy organization so that they have a context in which the artifacts will be demonstrated. Depending on what you are shopping for, for example a modeling tool or a simulation tool, that might be enough to give you an impression of the tool to be ready to score it.

Some organizations have a more peculiar stakeholder group and want to see their content in the vendor's tool ("custom data"). This should not be a big issue, since all tools will have some form of import and the artifacts are created relatively quickly. The exception to this is Process and Task Mining, which require an example analysis and will follow the steps of a "live" mining project, which is beyond the scope of a demonstration.

This phase will be done by the Solution Engineer and supported by the Account Executive who will provide available additional information in the form of brochures, White Papers, webinar recordings, and so on.

POC/POV

POC and POV stand for Proof of Concept and Proof of Value. The difference between the two is that the first has a more technical focus ("does this thing work for us?") and the latter has a more content focus ("does it do what we want to do with it?").

Proofs are an optional step and at this point should only be triggered if there is a serious interest from a buyer. It will involve a time commitment and engagement from both sides and multiple players, and will be run like a real project with defined outcomes and deliverables — but on a smaller scale. The underlying understanding is if the tool fulfills the objectives of the Proof, then the buyer will proceed to the next phase, the negotiations.

In a Proof, you will see not only a professional project management, but they typically also include enablement activities, and giving customers access to a separate technical environment that is provided by the vendor. If the content is sensitive data, there even might be an installation on the customer's premises. However, the main idea is that the buying team gets their hands on the tool, while being guided by experts from the vendor so that no misconceptions occur. A good vendor will also tell you if your requirements are too high, the tool can't do things, or how issues will be addressed in future releases of the tool.

Depending on the effort of the Proof project, be prepared that the vendor wants you to pay for his efforts. This also could include the deduction if the Proof cost if a customer buys the software, depending on the negation skills in the next step.

Negotiations

This phase starts with the decision on the buyer's side which tool they want (or the shortlist of not more than 2-3 tools), and might start with a formal RFP to be filled out, which might be just a formality at this point.

However, the vendor now will submit a formal offer and all the papers that need to be signed, and you will go into a negotiation phase. There are multiple levels the various sales people can give you discounts (for example, if you subscribe for a certain amount of time and not month-by-month, or if you have a progression in number of users or cases in the case of process mining), but there are limits to this and your Account Executive might have to chase multiple levels to come to an agreement that is acceptable for everyone.[35]

Depending on the type of contract (for example, did you buy a pre-packaged set of tools and seats, or do you have a custom enterprise setup?), the lawyers of both firms are involved, and they might have their special ideas to be included. One thing that I recommend negotiating in the first package, though, is the Master Services Agreement (MSA) that will rule all services that will be provided to you. In the future, you just have to agree on Statements of Work (SOW) for project phases, without having to agree on payment terms, processes, etc. every time.

Provisioning and base setup

Congratulations! You bought your tool. Now the real work begins, and the first step is that you get access to your instance (assuming that you are using a Software-as-a-Service/SaaS) provisioning. Otherwise, the installation and base setup is part of an implementation).

This typically happens when the Account Executive will trigger the vendor's cloud team to send the credentials for a superuser to the named customer person. That team will then start the conversations about the base setup, which typically includes the configuration of single-sign-on or of an email server so that the tool can send out notifications. Depending on your organization's (security) requirements, there might be additional technical things to be solved.

What you don't get in this step is a detailed training and the creation of content/structures or users in your system. These are things that happen in the implementation phase.

Implementation

The implementation phase is typically done with the help of the vendor's professional services organization (PS), or one of their partners, and you have been introduced to them in the sales phases before. If they are good professionals, they will follow the approach that I describe in this book and is presented in an overview in the "How to get your implementation off to a good start" chapter above ;-)

At this time, the sales team will disappear, and the project manager will be your main contact. However, good Account Executives will keep the contact with you and check-in occasionally so that you don't have a "cold" request if there is the need for more licenses.

The not-so-good Account Executives will see themselves as the single point of contact in their organization and demand that any request always go through them. Unfortunately, many topics in the implementation or later customer success phases are over their head, and they don't bring a lot of value to the conversation, or rather are delaying the solution of problems with their behavior. Make clear that you understand their role and that you'd rather work with the correct people in their organization directly.

Working with the services organization of a vendor/partner is a good thing because these people have seen basically all questions that you might have, and are involved or are part of the vendor's learning team. However, the challenge is that professional service organizations normally work on a project basis and when your project ends (successfully!), they move on and your contacts might be a dedicated Customer Success Manager or the Help Desk.

And, of course, the professional services organization will be happy to have additional projects with you. Those will be directly scoped and negotiated with the PS organization, and vendors do have different processes for service and license deals. As part of that, use the experienced PS resources that work with you to develop and update your roadmap and grab their knowledge of what has worked for other clients who have been in your shoes before.

Customer Success

If you are a large(r) customer or have a strategic importance for the vendor, you might get a Customer Success Manager (CSM) assigned. Typically, this person will be introduced during the negotiations phase so that they will observe what is going on in your program "as it happens".

The main objective of the CSM organization is to make you successful in reaching your business objectives that you have defined for your program in the Strategy phase (there is a separate chapter what to look for when developing a strategy below). As part of this, they will schedule regular meetings with you (typically on a monthly basis), and develop a client success plan together with you, which will also include vendor resources where applicable.

The CSM will be your main counterpart on the vendor side for the duration of the contract that you have closed, and I highly suggest building a good relationship with them. They can be the necessary "accelerator" in the vendor organization, which might be needed, and they can arrange meetings with the right resources and individuals to solve a problem, even if this is not "part of the standard process" — something that you want.
This also means that you should keep an overview of all open help desk tickets that you might have, and keep the CSM current on the progress (or lack thereof) of your program or the initial implementation.

Lastly, the CSMs in some organizations are also responsible for renewals when the contract period comes to an end. Make sure that you get the value from your investment and discuss your license (and services) needs with your CSM in advance, so that this does not become a "finance organization sends out an invoice" exercise, but both parties can track the completion of the customer success plan and have a good understanding if the assumptions that were made during the negotiations, such as the actual usage of the tool, are fulfilled or need to be adjusted.

Final thoughts

As you have seen, there are many people involved in the process and all of them operate under different conditions and have different, sometimes conflicting objectives. To avoid any problems or to de-escalate conflicts, I recommend to "play with open cards" once you have signed a contract.

In addition to this, you should clearly communicate what your expectations are and hold your end of any commitments and be a reliable partner through the sales, implementation, and customer success phases of your program. To get a head start on this, I recommend developing an initial strategy as described below and have clearly defined personas, before you even start the interaction with a vendor — clarity will be appreciated by everyone who is involved.

11

Getting Started Summary

Even though you do not need to look at technology as the first thing in your program, typically the first thought of an organization is "what tool do I need" to solve a problem.

The takeaways of this segment therefore are:

1. Everything is a process, and that is true for your architecture implementation as well. Even though you have four large areas (strategy, process/org governance, technical governance, and content creation/import), these follow their processes/methodology as well. These will be described in the following chapters so that you get your "cookbook" delivered with this book.[36]
2. Develop a clear understanding of what you want to deliver to your organization (EA strategy), for whom you do this (personas), and how you measure success. This will enable you to articulate clear requirements and help you choose the right tool and bring you through the buying process.
3. Communicate clearly and do not assume anything. Some people think that they might be considered "pedantic" or "controlling" when they track what they want, but I can tell you from many buying and delivery situations, that the clear communication that holds people responsible is appreciated by mature vendor organizations.

4. Select a tool with "room to grow in", which not only satisfies your current needs, but also has feature areas that you do not see in the next year or two. Your maturity will grow, and you might be at the time to look at advanced topics sooner than you might think.
5. Your tool should support the full process/solution lifecycle and not be a single-purpose tool. Even if it does not support one area natively, it should support standard exchange formats and have APIs that allow you to connect it to other tools that then become part of your tool landscape (for example an automation tool that can import a technical BPMN file from your design tool so that you accelerate your implementation and the overall design stays in the design database for consistency reasons).
6. The key feature of any tool is the meta model, and you should be able to change this when needed (e.g., to make the exchange of different tools happen), but stay with standard formats wherever you can. If you get pushback on this ("we are special" is something that we all heard from our mothers), keep your stand and insist on standard compliance. If you don't do this, you will be responsible for all adaptions and custom developments in the future, and that can get pricey pretty quickly and build up your technical debt. This will occupy your team just to "keep the lights on" instead of doing new things that let you mature your organization.
7. When selecting a tool, do not only look at features and functions, but rather take a look if there is an ecosystem around your tool and vendor. Are there user conferences that you can go to and exchange experiences with other users? Is there an ecosystem of partners around the technology who bring in their twists and extensions? Is there a proper learning offering which allows you to quickly onboard new users? You get the point — the success of your program is dependent on the people and not so much on the features of a tool.
8. When buying a tool, be aware that ultimately you, as the program manager of your architecture implementation, are the only constant in that game. However, during the different phases of that process you will

have different counterparts (like the Solution Engineer, Professional Service people, or a Customer Success Manager on the vendor side). Build good personal relationships with these — this is a small niche industry, and you might see yourself (maybe in conversed roles) again pretty soon.

Moreover, look out for online forums, LinkedIn, meetups, podcasts, etc. to see who is in this industry and connect with them.

III

Architecture Strategy

12

Architecture Strategy

"Strategy is having a plan while dealing with uncertainty" as I have learned from strategy consultants, and "just doing things that others do" is not a strategy, but a simple task ... one needs to eat to survive, but "eating" is not a strategy.

What's in this section?

This section covers the first of the four areas of your architecture implementation:

- The strategy that articulates clearly what you will and will not do.
- Proper communication to gain acceptance and bring your stakeholders and users on to the program (and convert the adversaries).
- Creating visibility not only into your architecture artifacts ("are we 5% or 85% done?" is a valid question), but also into your program and its progress.

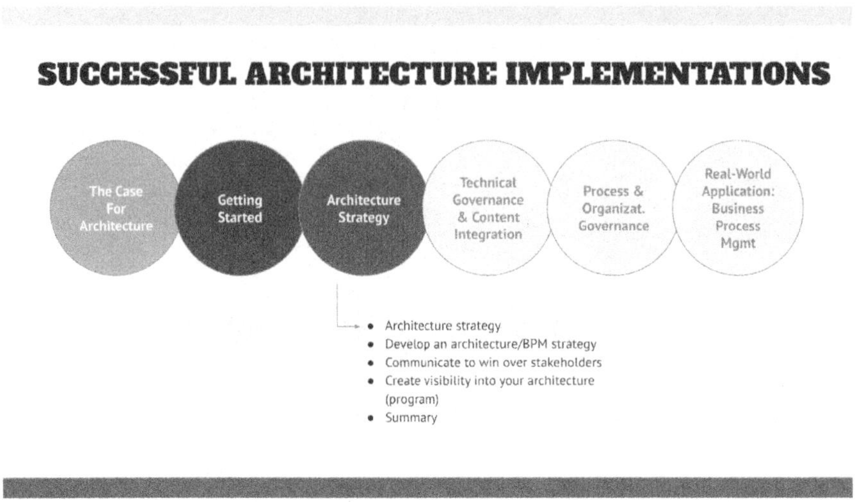

Fig. 12.1: Architecture strategy structure

13

Develop An Architecture/BPM Strategy

A few years ago, my younger son visited me in my home office during the day. He was in Middle School at that time, and I thought that I could ask him an adult question. So I asked: "What do you want to become when you're grown up"?

The result was not what I expected. His answer was: "I am too young to know that, but I know one thing for sure — not what you do!" Ouch!

Well, besides that, he thinks differently now, this is the situation that I see with numerous clients when they get started in their architecture journey. They have a gut feeling that they need to do "something" but can't articulate what they want and why they even look at tools.[37]

Why develop an architecture strategy?

I think the example above gives enough reasons to think about developing a strategy for your architecture program. But the other challenge that I see is that clients don't know how to develop that strategy. Let's take a look at what the components of a strategy are in this chapter, but first let's see why you want one:

A strategy is the formalized plan to deal with uncertainty. It is assembled of research and analysis, the development of a plan that is aligned with your organization's objectives, and the definition of how to reach those goals.

This will help your program (and your organization) to do the following:

- Act as a **guide** when things get unclear — the strategy articulates the "North Star" that everyone should work towards.
- It helps in **communicating the vision** to your team and stakeholders ("this is why we do this").
- Having clear objectives allows to **align all architecture developments** to these objectives, and this is also true for the implementation of the architecture program, which is "just one among many projects".
- A clear strategy allows you to **decide** what *really* is important, and helps to create the business case for the projects that you want to do as part of your program, or help other projects in their definition phase.
- **Adjust your strategy**, when you reached your initial goals or a milestone. When you have reached this, the uncertainty is gone (at least a bit), and you have to adjust your strategy — maybe not in the "big hairy goal" (your mission statement), but in the specific next steps that you are approaching. A strategy is not a "one and done" exercise, but will have to be refined over and over again.

What are the components of strategy development?

Now that we've defined that a strategy is not a monolithic thing, but an ongoing process, we can see what the components and the players in that process are.

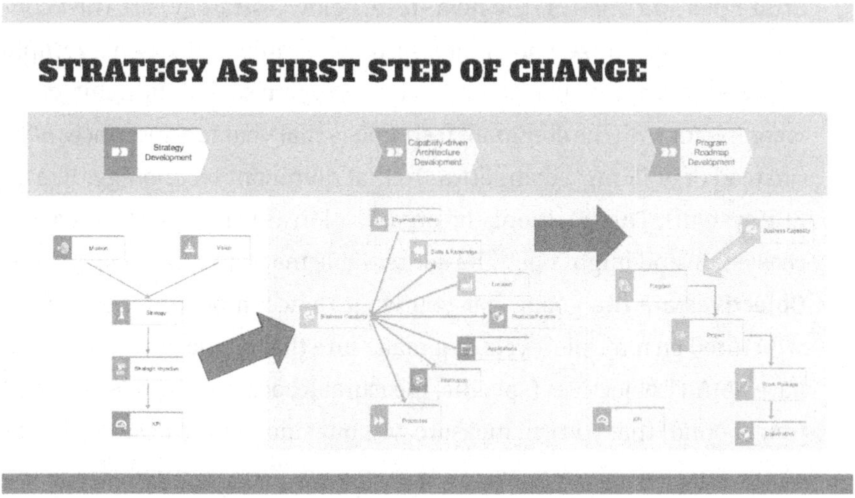

Fig. 13.1: Strategy as first step to implement change

A strategy should be documented in a logical structure, and what has been successful is a representation like in figure 13.1 above:

- The **Mission** for your efforts. What is the target state that you want to reach? How does this improve your organization? Why do you want to implement architecture? The mission statement is a short sentence (or two, but not more) that brings things to the point.
  One example could be: "We deliver better products to our customers by having a solid architecture that reuses shared components. It also defines how we develop our products and improve our internal processes and quality".
- The **Vision** of your program: What is the qualitative change that you will bring? What does the future look like? What will your stakeholders (customers and internal stakeholders, as well as [external] partners) be able to do better?
  This can be a lofty statement ("Just do it" is what Nike distilled theirs to, or think of the "Think different" manifesto that Apple developed after Steve Jobs' return to the company).[38]

- **Strategies** and **Tactics** (the next-level below a strategy) are the "containers" to organize your plans into actionable items. An example strategy can be to restructure a product program so that its architecture is compliant with the dominant technology that your tool interfaces with (instead of building a competitor to that dominant technology, like an ERP system). This positions the organization as part of the ERP system ecosystem and might open the addressable market to more people.
- **Objectives** are the goals that you want to accomplish. They can be structured on multiple levels, but make sure that on the lowest level you have SMART objectives (specific, measurable, achievable, relevant, and time-bound) that you can measure without ambiguity. These objectives can be internal ("How do we want to change?") or externally triggered ("What do we want to accomplish in the market?").
  An example of an objective for your program could be that you have implemented your architecture tool and onboarded three programs / two org units by the end of the current fiscal year.
- Lastly, there are the **Key Performance Indicators** (KPIs), which will measure the accomplishment of the objectives. To stay in our example for the objective, you might want to measure the number of users and their certifications. Or the amount of existing content from the two org units that have been reviewed, cleansed, and transferred into your repository.
  These KPIs then will be implemented by the projects that are part of your strategies or tactics and close the loop to the strategy development.

As mentioned before, you want to collect your accomplishments (or failures) and adjust your strategy accordingly.

Steps for developing your strategy

Developing a strategy for your program can be broken down into two parts — the definition of objectives and the design of the implementation that will result in projects, as figure 13.1 above shows.

Figure 13.2 below shows the detailed steps for the strategy development itself:

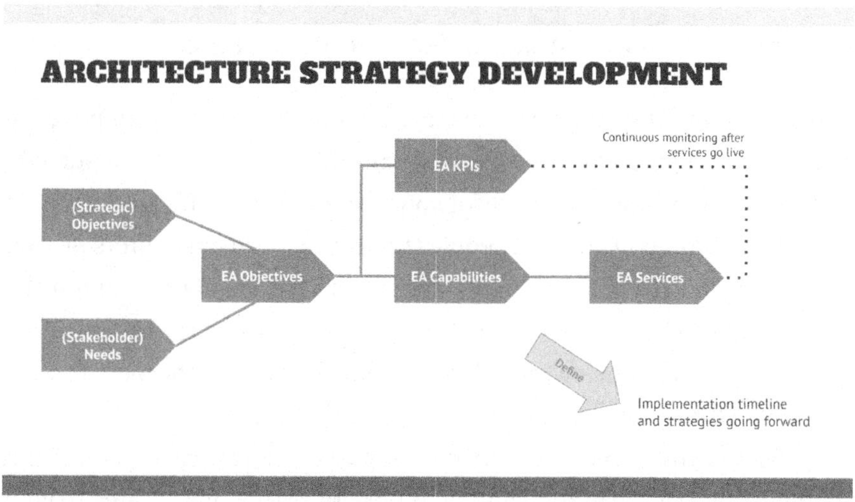

Fig. 13.2: Steps to develop a strategy

Identify the strategic objectives of your organization

The first step is to do some research and try to find the strategy definition for your organization. What is the positioning of your company on the market, and where shall it move to? What are things that are currently measured (for example, Annual Recurring Revenue (ARR) as a sales KPI, compared to traditional license sales and maintenance, and renewals in a software company).

Unfortunately, many organizations that I've worked in didn't follow the process above when developing their strategy, so you might find some lofty presentations, but not "hard-wired" objectives. This leaves room for interpretation, which is not good, so you will have to verify what you found with the authors of the presentations.

However, collect whatever you can get and see which of those objectives can be supported by your ideas of implementing an architecture practice in your organization.

Identify the relevant stakeholders and their needs

The second step of research is to identify the stakeholders for your program. These are more than just your bosses who will have to open the purse to implement your ideas. Look out for your potential users (future personas), directly or indirectly affected people (for example, business units or your risk/IT organizations), external stakeholders like regulators, and influencers that can be your multiplicators going forward. In the organization governance chapter below, you find a graphic that shows the main roles in an architecture program as a start for your first list of people.

For each of these roles, identify the people who hold them and reach out to them. You want to understand their objectives (not only related to your program) and how they get measured/what success means for them. Also ask them for their timelines in projects.

Collect this information in a spreadsheet or database, and use that information to identify the "what's in it for them" — how your program can support them and make them more successful, for example, by accelerating their roll-out or avoiding custom development when you already know that a similar technology already exists in the organization, and you can "connect the dots" for them. This will also help you to identify the services that your program will provide to them (see below).

The second thing you should do is collecting the timing information that you might have heard on a calendar. This will help you in planning specific measures, such as onboarding a different group because of the stakeholder that you just spoke with.

A last thing you might want to do — but never share with your stakeholders — is to create an influence map.[39] This will show the strength of influence between the different individuals and help you to identify who will be a supporter and who can be a distractor.

While doing this, also determine how the stakeholder feels about your program: are they excited, supportive, neutral, or opposing your ideas. This will determine your approach for communication with them, and you might want to try to convert an adversary to a "believer" who will be a great asset for you going forward. Those might not necessarily be the ones who speak the loudest, but the silent people in the back who try to stay out of the limelight.

Define the objectives for your EA program

Now that you have collected the necessary input from your organization and the stakeholders, it is time to add your objectives to the list and define the objectives for the architecture program.

This is typically done in a tree-like decomposition (similar to the capabilities or services shown below) and you should make sure that each objective can be related to or supports one or more objectives that you have heard. Moreover, take a look at the timeline that you created in the previous step when it comes to the timing of the objectives and the potential priority.

Occasionally, that relationship might not be created, but focus on identifying conflicting or overlapping objectives that you might discover. Try to resolve these with the stakeholders (maybe you simply misunderstood them) and flag them as a risk if you cannot resolve them.

Like in every project management system, you should have a running list of program risks that you need to classify, develop potential ways to mitigate them, and put them on a timeline. This list should be reviewed on a regular (and frequent) schedule. If it involves personal animosities, try to have someone else solve them on behalf of the program!

Once you have assembled a first version of the objective tree and a potential timeline, socialize it with the stakeholders and your team. You want to gather as much feedback as you can, and also update your stakeholder maps from above.

Identify the capabilities that you have to set up

Once you know what you want, this is an easy exercise. Take an architecture capability model and identify which capabilities you need to put in place (or uplift) in which sequence.

I have developed the capability model below while working in multiple organizations. In projects, I printed it on a poster-sized sheet of paper (or "ghetto fabulously" taped together dozens of letter-sized sheets of paper) and then marked which capabilities shall be grouped into different waves of implementation.

The capability model follows the five main "buckets" described in the "Which capabilities do I need to build my practice?" chapter above, and the waves typically followed the sequence of the questions that the capability groups answer.

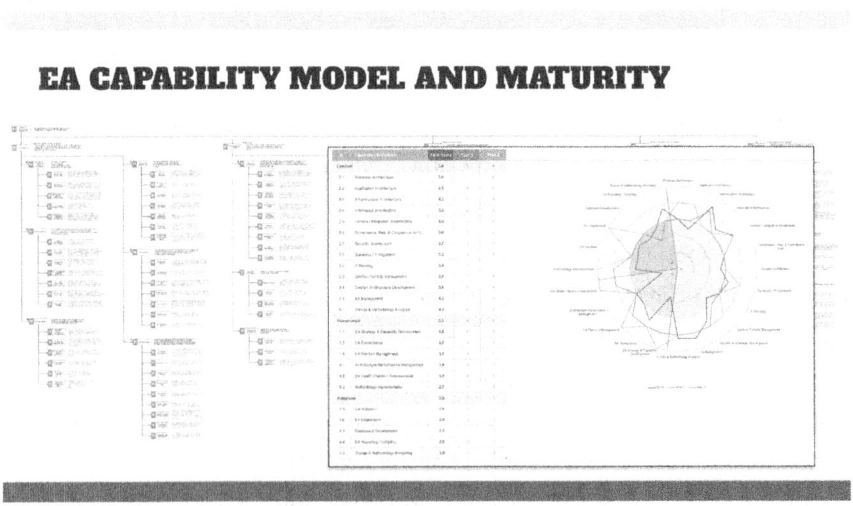

Fig. 13.3: Architecture capability model and maturity assessment

Another thing that you want to use the overall capability model for is a regular assessment of the architecture maturity of the different groups in your program. Not every group in your program will have to develop all capabilities, so you might have to split it into different surveys, but what you want is to assess the maturity following the CMMI maturity levels:[40]

1. **Initial**: Processes are unpredictable, poorly controlled, and reactive
2. **Managed**: Processes are managed on the project level and are frequently reactive
3. **Defined**: Processes are well-characterized and well-understood
4. **Quantitatively Managed**: Processes are measured and controlled
5. **Optimizing**: Processes are stable and flexible

The different capabilities are then mapped to the three areas of successful implementations (also described in the same chapter above): Content — Governance — Adoption, and can be displayed in a radar chart. The idea is that you reach at least a level 3 rating ("Defined") as quickly as you can, so that actions will be aligned.

One thing that I did in the past was to speak with my clients about their ambitions for the next year and the year after that. This helped to guide the adjustment of the strategy for these time frames, and made it clear where the focus would be going forward.

Derive the services to your organization that your program will provide

The next step is to define the services that you intend to provide to your organization. The easiest way to understand services in this context is to see them as "Capabilities that have an SLA attached to them." Which means that you define how well you will bring the capabilities into reality.

In the "Changing role of architects" chapter above, you have already seen which services the different groups and roles provide and consume, and a Center of Excellence or an Architecture Review Board are two of the more prominent examples of service providers. In the following governance section, we will talk more about how to structure and set up these organizations, but to nobody's surprise — I hope — this will be done in a process-oriented fashion.

Typically, I like to define the services in multiple categories that are shown as an example below:

- **Foundational services**: these are services that are needed to "keep the lights on" for the architecture program, like managing the architecture tool or defining standards, and supporting users with training and other activities.
- **Baseline architecture services** include content work, such as creating and maintaining reference frameworks, or providing expertise and man-power for things like application rationalization or process modeling.
- **Additional services** — this can be multiple groups that help individual programs or more generic solution development support, as they are described in the Process/Solution Lifecycle chapter. They also can include coordination tasks, as well as laying architectural runway for larger solution programs. The architecture group can be "grease" between those large programs, making sure that what these programs develop will fit into the overall strategy of the organization.

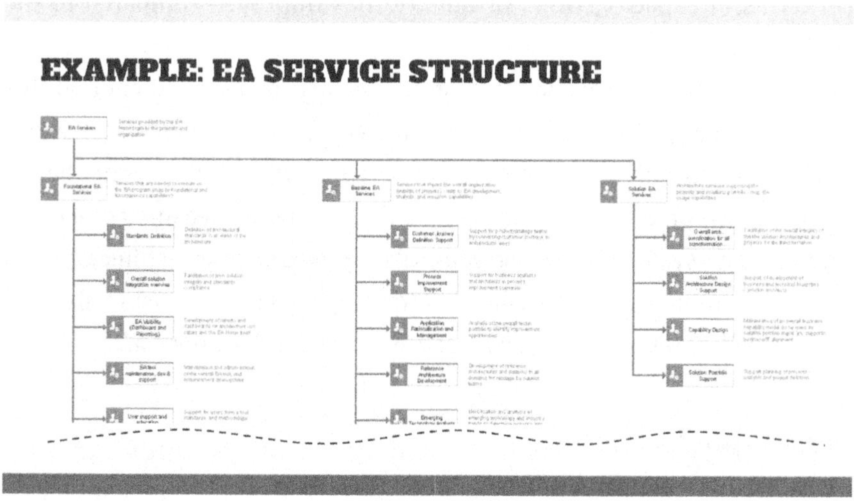

Fig. 13.4: Example architecture services

Like all outcomes of the strategy development steps, the service structure needs to be updated and is not a "one and done" exercise. Plan regular checks of the quality of the services that you provide and use that input when it comes to budget planning for the next fiscal year. You also want to align this to the maturity assessment of the architecture capabilities because you might need to do more on that front before you can deliver a service to users. Users might not be interested in how you do things as long as they get what they want, and you might do things for "the greater good" (or to avoid that they don't fall on your feet later).[41]

Identify the KPIs to measure success

The last step in the Strategy Development process is to identify the KPIs that will measure the degree of accomplishment of your objectives. One thing that you should have in mind when defining the KPIs is that you identify KPIs that have a business impact. For example, do not make the number of training participants in a system implementation the KPI for implementation success, but rather see if the performance of your process (in terms of capacity, time, or quality) has improved compared to the "before" status.

When doing this, think about a systemic measurement — do not do this as a manual exercise because it will "fall asleep" when the person who is initially tasked with that moves on, or the reporting of a KPI does not have a direct impact on the people who collect them. Having measures that come directly from systems — maybe your sales system or your architecture tool — via an automated report or dashboard feed, will be more sustainable and avoids the impression that you measure things for their own sake.

Now what?

At this point, you have assembled all the ingredients of a proper architecture strategy, and the only task is to prepare the meal and gather the feedback of your stakeholders.

But in all seriousness, the next step is to take all the different deliverables and group them into logical packages that will represent your strategies and tactics. Those will then contain the necessary projects that will implement them. Since you already know what the outcomes should be and how you measure them, this exercise should be an easy one.

However, do not forget to do two critical steps:

- **Articulate your strategy in a wiki** or other always accessible tool (and link to more detailed analysis and supporting material where appropriate). Do not make the mistake of capturing it in a document or presentation that will be soon forgotten on a shared drive or sinks further down in someone's inbox.
  Your strategy shall be the lighthouse for your program, and everything you do should be logically traced back to what you've defined.
- Once you have a draft, go and **socialize the whole package** — you have already captured feedback on individual items, but now the stakeholders will see "the full package" in one place and might have additional feedback for you that might be not as narrowly focused as you might

have received feedback for earlier.

The important thing is that you can address and resolve any conflicts so that you get the buy-in from other programs, for example, and are not considered someone who makes their life harder, but rather helps them in achieving their success.

Lastly, you need to think about how to introduce the changes, while building your systems to be able to measure the success of your strategy. The next two chapters will cover those two aspects in more detail.

14

Communicate To Win Over Your Stakeholders

Establishing an architecture practice is not only establishing rules, bringing in a tool, and creating content. It is first and foremost an exercise in communicating and convincing others that building this capability is a good idea. There are just a few scenarios where it is mandated that you build, for example, a process inventory and in all other scenarios you depend on the good will of others if you want to succeed.

Change is hard

Implementing changes is hard and complex. It is not done by just bringing in new technology, but you also have to adapt processes, define measurements, enable people, negotiate with others, and so on. And the hardest thing is to change people — or, as I like to say over and over again, it doesn't matter how many wallpapers [of processes] you create and how many boxes [of IT] you put into the basement. If the person in front of the screen does not want to change, you lose.

This is not new and the organizational change management practice has long figured out what the stages of change are, as shown in figure 14.1 below:

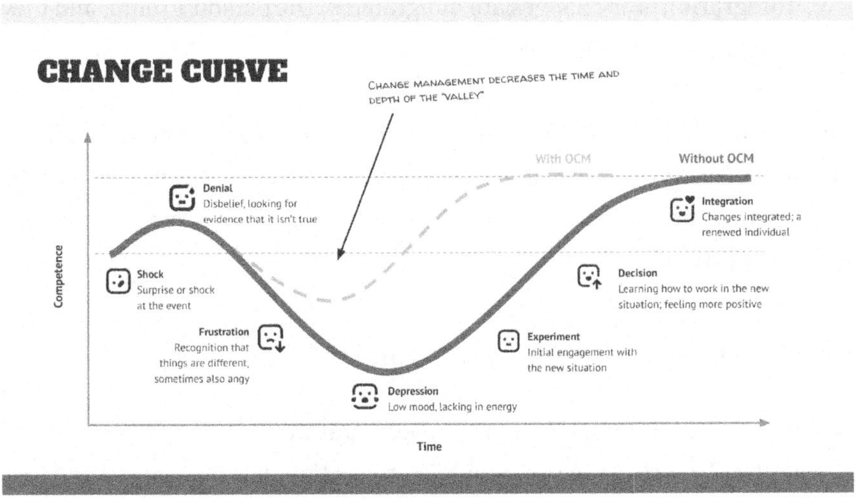

Fig. 14.1: Change curve

The idea of every change is to bring the performance of your organization to a higher level. But then you see people reacting in unexpected ways — you typically see the following stages:

1. **Shock**: for most people, major changes come as a surprise and a normal reaction is to be surprised or in more serious situations fear.
2. **Denial**: as a reaction to the event, the productivity might even rise a bit because people might want to show that there is no change needed. They might seek evidence that the announced change is not true.
3. **Frustration**: it sinks in that the change is coming and things are different going forward. Some people might even react angrily.
4. **Depression**: this is the low point in the change curve. There is a low morale and mood, and a lack of energy might show.
5. **Experiment**: after going through the low point of the curve, some interest shows up and an initial engagement with the new situation starts ("maybe there is something in this?").
6. **Decision**: people learn how to work in the new situation and start feeling more positive.

7. **Integration**: the changes are integrated in the person's mind, and they perform on a higher level.

There is no way to avoid the change curve, but you can avoid the depth of it and also shorten the time to go through the different steps by using communication and enablement. This is also true for your architecture implementation.

Impact on your implementation

So, what does that mean for your implementation?

Well, the first thing is that you have to realize that it is more than just handing out licenses to the tool that you have brought in (in case you haven't noticed). You rather have three things that you need to balance in your program that are shown in figure 14.2 below.

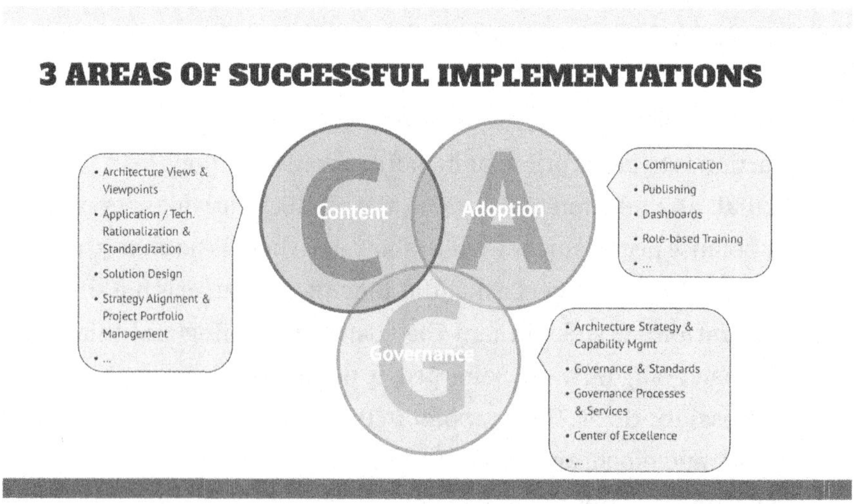

Fig. 14.2: Three areas of successful architecture implementations

I wish I could give you a cookie cutter advice on to which degree each of the three areas need to be balanced, but this is highly dependent on your organization. Or more specifically, the culture and people in your organization and how they react to change.

But let's take a look at the three areas first:

- The first area is "**Content**" — this includes all the artifacts and use cases that you have identified in your strategy. It can be as broad as a use case "we want to use it for Solution Design and this is how we have to change our Software Development Lifecycle (SDLC) to make it fit," or as little as "we want to standardize on one process notation firm-wide so that we speak with the same language". [42]
  However, you can already see the potential conflicts when you read these two statements ... people might love what they do today and not like your ideas.
- The second area that affects people is the "**Governance**." How do you want to "run the show?" What are the standards that you need to implement, and how do you get things done?
  Those need to be explicitly stated and be available for everyone to access and give input on. Governance is not a static thing, but — as in all governments — a balance of "checks and balances" in itself. We will talk more about this in the next sections.
- The third area is "**Adoption**" and this is where you are right now: the identification of the changes, the needed measures to address them, and the enablement of the involved people.[43]

By realizing that your success starts and fails with the proper adoption, the topic of frequent and constant communication becomes apparent. When adapting a good communication approach, you will be able to win over stakeholders (purse holders and end users) and be on time to get the attention, support, and budget that you need to succeed.

What are the steps for developing good communication?

There are a bunch of approaches to Organizational Change Management (OCM) and I highly recommend studying those, and involving the OCM people of your organization if you have those. They know. However, I like to focus on communication in this chapter to make things a bit easier to apply. Here are the steps to take:

1. Determine objectives that you've identified in the strategy phase, basically answering **the "why" and "how"** questions (objectives and capabilities/services). These two topics will be the main content of your communication effort in the beginning.
2. Then identify **stakeholders** and reach out to them to not only confirm their objectives, but also to understand their motivations and "what's in it for them" — the triggers that might move them to take over a larger role in the effort ... you will need sponsors for your program.
   The stakeholders can also give you an idea of the number of people involved in their areas of responsibility, which will have a practical impact on the measures you will plan. I suggest capturing these in a table and using it as a reference, and to confirm the items in it with the stakeholders regularly.
3. Once you have understood the landscape of your communication audience, create a **communication strategy** document and socialize it. This document should include the main messages, stakeholders/sponsors, communication and feedback channels, the addressed community and how to build/support one, and the KPI for measuring success.
   This document can act as a "lighthouse" and blueprint for different waves of communication activities.
4. The **communication plan** is a document[44] that will then list the individual activities that you want to roll out and how you measure their success, for example by looking at the opening rates for newsletters or participation in forums/channels.
5. Then create the actual **communication items** per your plan and **mea-**

sure their effectiveness. This could be electronic communications (as if we don't receive enough emails already) or analog communications — think of newsletters, Slack postings, road shows, print material, or (as I had in one project) an installation: we placed an 8 ft (2.44 m) tall acrylic tube in the lobby of the building and filled it with pebbles to show the migrated accounts in a system implementation for a national telco provider.

Whatever you do, think about the efforts to maintain and reuse the material, and also if this fits into your communication receiver's mind set. If everyone works remotely, the installation idea might not be very effective.

6. Set up a **program-wide measurement** system for your communication. You want to have "hard data" of what has worked in which part of your organization and what has not. What you definitely would like to avoid is to lose an org unit because their local culture is not responsive to your efforts, and they get stuck in the valley of despair of the change curve. Since bad news travels faster than good news, such an org unit could bring your whole program into danger.

   You also want to do this for your stakeholders and have regular check-ins with them to see if you are still on the same page and on track to reaching your goals. You need to constantly show the business value to them so that you can get their support for future phases of your program.

7. And lastly, think about what happens with all the good information that will be created in your implementation program for the time when the initial implementation is done, and you go into a "business as usual" steady state. Think about how your organization does **Knowledge Management** and use the tools that these folks already provide.

   If you don't have any of this, I recommend installing a wiki or internal website as a minimum to provide support guides, procedures, tips and tricks, and communications. You should also provide a space for conversations — either in person ("office hours") or electronically (forums, Slack channels).

And lastly, please keep in mind the effectiveness of human connection and meeting in person: when you reach a milestone, go and celebrate!

15

Create Visibility Into Your Architecture (Program)

One of the main challenges in an architecture program is the lack of visibility. Even if you have rules in place that shall help, such as mandatory attributes for models that you can report on, it is difficult to answer the question "what is going on".

And you will get asked that question often — either by yourself or a stakeholder.

There are many ways that contribute to an answer: reports, dashboards, project status information, etc. But you will notice quickly that you can't answer all questions with your architecture platform alone.

Let's take a look at the TOGAF ADM cycle, that we've met as a framework in an earlier chapter. In the first phases, it requires you to describe your architecture (Phases A-D) in figure 15.1 below. Then you look at all the different ideas that you might encounter and feed them into a pool of requirements to be able to look at them at some point. Then you can decide which ones to implement in a solution, and plan the implementation (Phases E and F). And lastly you implement, execute, and measure the running architecture ... something that isn't even included in the ADM cycle.

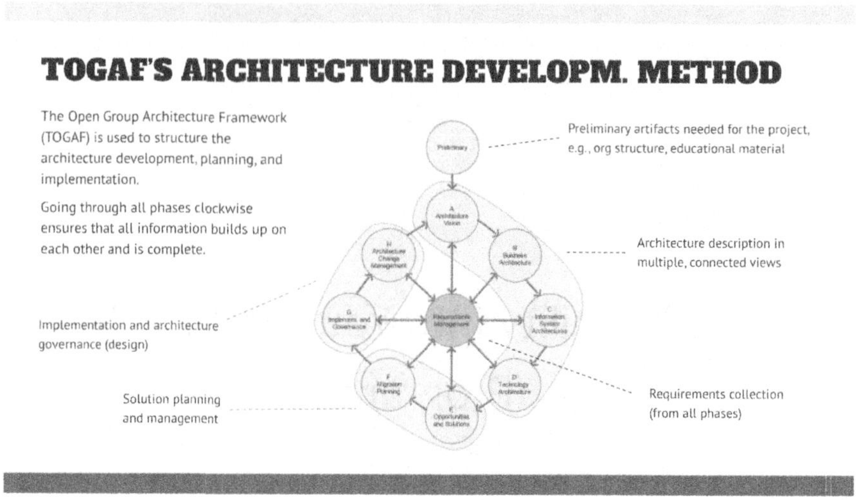

Fig 15.1: TOGAF Architecture Development Method

But how do you know what to look for? The status of an individual architecture or artifact? For an overall status of completion of your data collection? The implementation status of the projects or the benefit realization of your previous improvement proposals/business cases?

The answer to this is that you will need dashboards that are fed by different systems and — and this is the important part — allow you to take action and not just display pretty charts without context.

To answer these questions and create dashboards that will answer them, you can follow the process below.

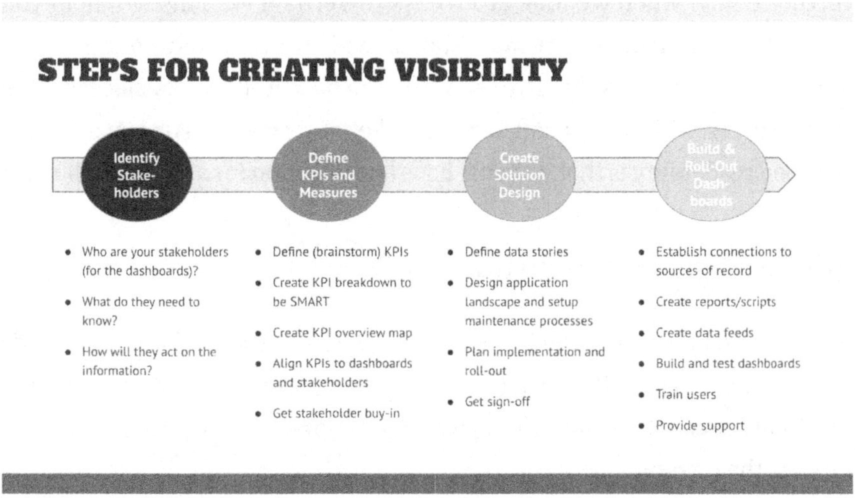

Fig. 15.2: Steps for creating visibility

Identify stakeholders

The big mistake that I see in dashboards, which makes them unreadable and challenging to understand, is that people want to build one dashboard for everything.

The better way is to **review and understand the different stakeholders** that you would like to provide information for. What do they need to know? What are the questions that are relevant for them?

A good start is the stakeholder analysis that you did as part of the strategy development, but now you will have to open your perspective a bit more. You need to include other stakeholders, like collaborators, external partners, IT operators, etc. so that you can tweak the visualization in a way that makes sense to them.

Lastly, you have to define **which actions** these stakeholders will trigger based on the information that you display in your dashboard. Your dashboard now transforms from a "pretty to look at" overview to a critical application in a process that they have to do (and we will discuss processes further below when we look at Process Governance). This will help you **identify if you need one or more dashboards for a stakeholder** group — and in almost all cases that I've seen, you will have multiple dashboards (the single screen with widgets on it) for a stakeholder group. The key principle for good readability is that you only have one question for a single dashboard.

Define KPIs and measures

Now that you have identified your stakeholders and their information needs / questions in a long list (and different stakeholders might have the same question), it comes to **identifying the Key Performance Indicators** that will answer the question.

One way to visualize this is in the form of a KPI Tree diagram that decomposes the KPI over multiple levels down to a level of detail that you can measure it — how many are there? How long does it take? And so on. The lowest level must not have any ambiguity or interpretation wiggle room. The lowest level KPIs can also be reused, as you can see in the tree on the left-hand side of figure 15.3 below, where the second KPI from the left on the lowest level "feeds" two higher-level KPIs. There is no structural "guardrail" on KPI Trees — decompose them to the level that you require until you reach the lowest level that cannot be further decomposed.

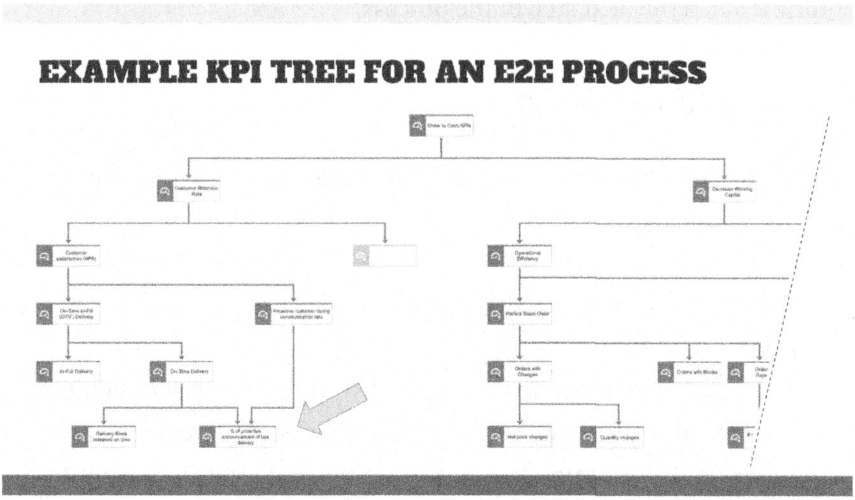

Fig. 15.3: Example KPI breakdown for a process (Order-to-Cash)

And, of course, you can make your KPI Tree prettier by showing the descriptions and adding the questions to the higher-level KPI objects. You can also add the percentage that goes into the calculation of the higher-level KPI to indicate which one counts more (e.g., 70% to 30%).

The next step then is to **align the KPIs and questions to stakeholders**, and a simple table will do the trick. This will show you if there are overlaps between the stakeholders. You can also do a rationalization if you have all the KPIs/measures that you need when you look at the tree with the perspective of a different stakeholder.

However, don't overthink it. You don't have to have thesis-level granularity in your measurements, but catch the most important one. It is better to have an answer out for your stakeholders so that they can act, than trying to be perfect — you can always tweak them later after you have satisfied the main information needs ... and sometimes people come up with their ideas when they see a "version 1" of their dashboard.

Lastly, **socialize the KPI Tree and the matrix** with your identified stakeholders and get their immediate feedback. They can also tell you where you can find the systems for the lowest-level measures, and you can use this information in the next step of the process.

Create solution design

Now comes the fun part — creating a solution design. This typically centers around these areas:

- Identify the **processes and tasks** in which your dashboards will be used.[45]
- Define the **data story** that is needed to support the individual task. What is the sequence of information that is needed? What is the most important information? How does a stakeholder drill down into the data? Figure 15.4 below shows a dashboard from a process mining analysis as an example.
- Design the **application landscape** and **data models** for your dashboards. What are the sources? What are the file formats that you get from them? How do you get access to the source systems, and how do you automate the data pull? What objects and attributes do you require?
  This will include many conversations and data transformations, like consolidation of multiple sources into one feed, and you might encounter some technical, but also human resistance when you ask for the data. Maybe someone has to write a report "just for you" in his systems that he wouldn't have done otherwise (aka, he's doing you a favor).
  Communicate clearly what you want and how you will use it, and invite the people with concerns to be a reviewer of your solution architecture documentation to provide feedback and show them / provide them with access to the final dashboards.
- Document the **operations of the interfaces**. Ideally, you will be able to automate everything, but occasionally, you need a human to trigger a report or post a file somewhere for you to grab. It needs to be clear what needs to be done by whom, and what the frequencies/service levels for this are.
  Socialize this with the involved parties and get their buy-in (and think

about what you can do for them in return).

- **Plan the implementation and the roll-out**. What has to be done? Where do you do development and testing? How do you do versioning and publishing? What training is needed for the stakeholders? And so on. Assemble this in a document or on a wiki and communicate clearly your timeline, and what is to be expected to be done by whom. Track the activities and results ... just like in any other project.
- Once you have created your solution design, **get sign-off** from your stakeholders and the involved parties.

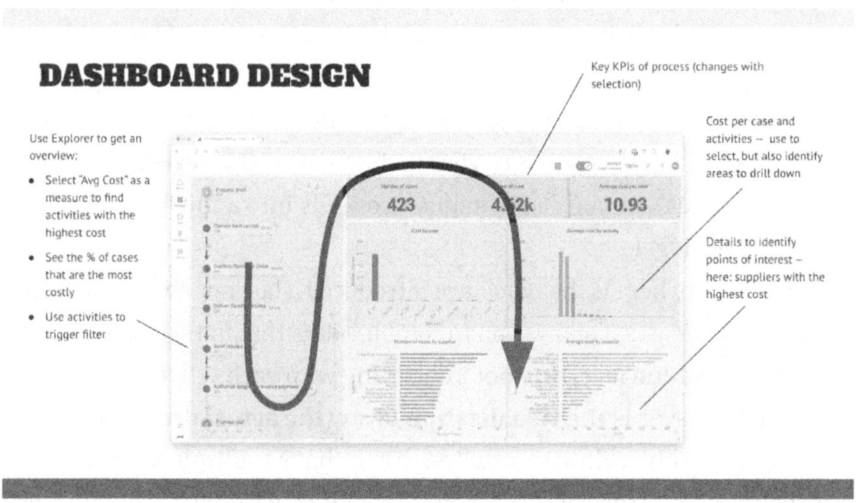

Fig. 15.4: Example dashboard telling the story of this analysis step and guiding the user

Build and roll-out dashboards (and reports)

The last step of the process is the building and roll-out of the dashboards.

Depending on the dashboarding software that you use, things might be slightly different, but in general, you will see a three tiered setup as shown in figure 15.5 below:

1. **Data sources** — this can be structured (tables) or unstructured data (text, pictures). Ideally, you want to have structured data to be able to easily transform it, but sporadically, you just pass through unstructured data "as-is" (like map visualizations or feedback from idea management systems).
2. **Data feeds** — typically, you require one data feed for a widget in the dashboard. This level now might take one or many sources and transform the data into one table that will feed the widget.
   It makes sense to think about the various data feeds before you create them (and maybe create some data models as part of your solution design, above). You can reuse data feeds for other widgets by adding additional columns, and you just use some for widget 1 and others for widget 2 (which will reduce your maintenance effort when you need to do changes), or you can combine two feeds into a third one that then feeds a widget.
   The beauty here is that you have introduced a layer of abstraction from the source which allows you to combine data that in the real world (on an application level) are not interfacing with each other.
3. The last layer is the **visualization layer (the actual dashboards)**. All major dashboarding tools allow you to have multiple tabs and even have not only the widgets on one tab interact with each other, but also allow configuring the interaction of tabs with each other (for example when you filter something).
   Now come all the initial thoughts together: What are the dashboards "apps" that you have to create for a stakeholder group (remember: you want to have one or more apps per stakeholder? Try to avoid mixing them and build "universal dashboards" — be specific). What are the tabs of the dashboards, representing only one question at a time for this stakeholder? How do you implement the data story on a single dashboard to answer the question in the most effective way? And what

is the action that a stakeholder shall be able to take when using the dashboard — can you include buttons that trigger workflows or reports (or any other action that the process in which the dashboard is used requires)?

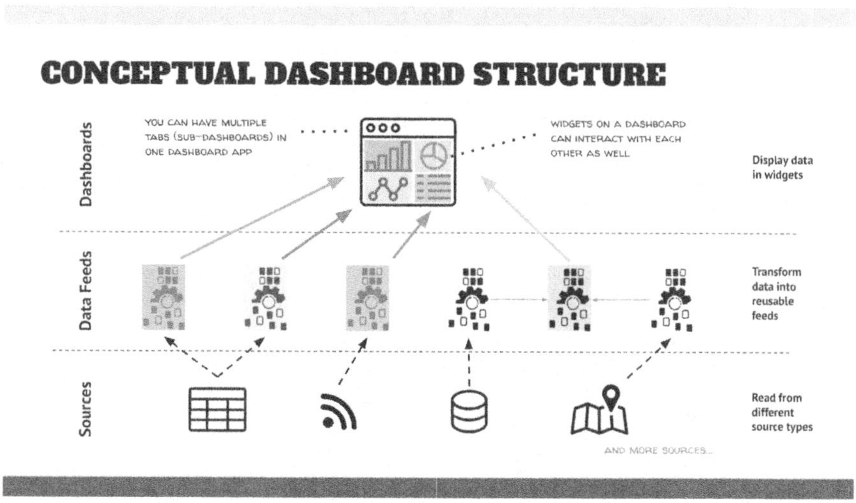

Fig. 15.5: Conceptual dashboard structure

Once you have built the dashboards you need to test them, and this should be run like any other IT development (multiple environments, testing, proper roll-out and training, etc.). In the next section, I will talk about how to do library import of objects from systems of truth and describe some principles and activities in more detail.

Overall, you want to treat your dashboards like any other custom application that your organization builds and maintains. You want to make sure that it gets registered in the application catalog so that it will be managed and maintained even after you've moved on to other activities and the IT "business as usual" kicks in.

Lastly, you want to train your stakeholders (which could be a short demonstration or video) and provide support for the dashboards. The knowledge base that we discussed in the previous chapter can be a good resource for this too.

16

Architecture Strategy Summary

"Strategy" seems to be that nebulous term that is never well-defined or communicated clearly, so that groups can align their efforts and activities to it. This segment should have given you enough input to avoid this scenario.

The key takeaways are:

1. Developing a strategy for your architecture program is an act of diplomacy while also having your interest in focus. It includes the juggling of multiple related, and sometimes controversial things and forming it into a logical whole that will allow everyone to understand who you will be "when you're grown up".
   Hint: this will never happen, so you will have to update your strategy regularly to adapt to a changing world — people leave, projects get completed successfully or fail, or the business environment changes for your organization. Include some flexibility in your strategic thinking and avoid being dogmatic ... this is one characteristic that nobody likes in my experience, and you will be excluded from the action going forward — either openly or behind your back. The result will be the same, your program will fail, not get the necessary funding, or the people assigned to your efforts "vote with their feet" and leave you stranded.
2. Besides updating your strategy, keep a tab on assessing how well you do

things. This can be regular capability maturity assessments or surveys on the satisfaction of your "customers" (aka, the groups that you provide architectural services to). Take this seriously and also show some flexibility — in most cases the feedback is not meant personally to you, but people want to help to improve things. It is just that the medicine that the doctor ordered for you might not be the best tasting...

3. People do not want to change. And changing people is the hardest thing to do. Understand the phases in the change curve, and understand that everyone goes through that curve individually and as a group. Address that situation with empathy, but also stick to your guns when it comes to implementing the change — the idea is that you make the organization better performing, and eventually people will realize this (but keep an eye on stakeholders who might be an adversary — you want to convert them to advocates or get rid of them).
4. Organizational Change Management (OCM) is your key advantage when it comes to implementing your program. Being able to communicate clearly and structurally sound (as shown in your communication strategy and plan) is crucial. But also keep in mind that you need to address and react to the feedback that you might receive — see above. In these situations, swallow your pride and try to understand what needs to be changed to accommodate the feedback and improve.
   Remember: you are most likely setting up your architecture program to help others. I have not seen an organization creating this capability "just to have one".
   The second (and third) part of OCM is organizational design and enablement, and we will talk about those in the Process and Organizational Governance section of this book.
5. Focus on creating a structured approach to visibility, and avoid manual, individual heroics whenever possible. You need to be able to communicate the value that you bring to the organization instantly. The best way of doing this is to clearly understand your objectives and communicate the benefits of your program's work in dashboards that are relevant for your stakeholders. Show how your team's work positively influenced

the overall outcome, but also each individual project. And remember: your program is "just one of many" and needs to justify why you want a certain budget or resources going forward.

6. Lastly, keep in mind that you are building a knowledge base and establish methods for everyone in your organization. Avoid securing the "employability" of your team and unnecessarily keeping the architecture work an expert task. This will just create bottlenecks in delivering outcomes — something that is on your list to remove, right? Architecture needs to be freely accessible to everyone because you might not have thought of all potential use cases when you start out with it. The better run programs I have worked on took the attitude that "knowledge is the only resource that does not get diminished when shared" and they took this into account when it came to user licenses, communication, and enablement. You never know if that person in that obscure department that you don't know well, might be appointed to lead the next large program that secures the organization's success in the future — you want to be part of that party!

IV

Technical Governance And Content Integration

17

Technical Governance And Content Integration

Nothing says "old school enterprise architect" more than listening to nerds talking about methods, meta models, and the proper usage of the tools. And we have already used some (digital) ink above when we spoke about frameworks, methods, notations, etc. Now is the time to put this into practice and in your tool.

Even though this is just a small part of an architect's work, we will spend some time in nerd land as well.

What's in this section?

I will give you an introduction into how to set up your tools and provide an environment for your people so that they can create the architectural artifacts that you need in your organization. This covers:

- The approach to set up technical governance (data model, meta model, and so on) and configure your tool, which will be shown using the ARIS tool by Software AG.
- Then we look at the setup of multiple databases in your environment to establish a release process. But also ensure the integrity of your

information when it comes to imports and synchronizations with other tools, as well as compliance by maintaining an archive.

- In the last chapter of this section, we investigate what it takes to integrate your tool with other systems of truth, and how this application landscape for architecture work within our organization can look like.

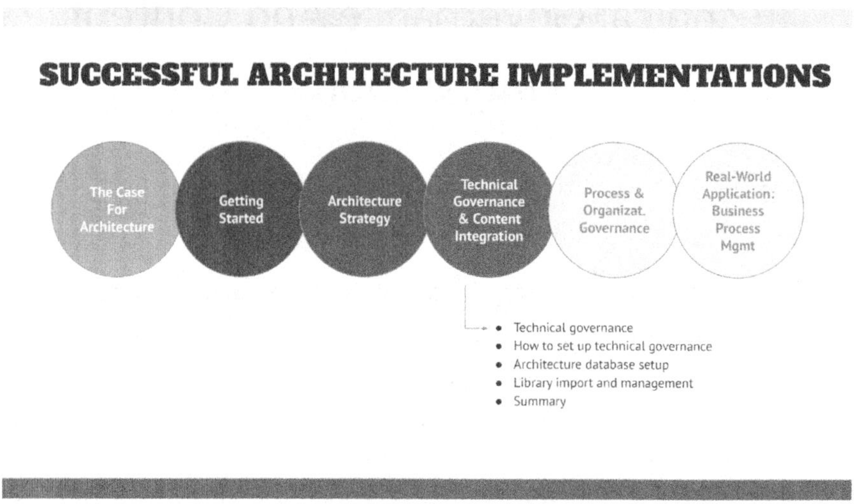

Fig. 17.1: Technical governance structure

18

How To Set Up Technical Governance

Defining the Technical Governance as part of your implementation seems to be a quick exercise – why should you create data models and a meta model when configuring the method in your architecture tool? Isn't that an effort that one can save and just run a report that spits out the details of the configuration when needed?

I disagree with this approach, and will show in this chapter how creating these two types of models will help you identify what needs to go into the configuration, and also avoid multiple "add-on" releases of the configuration. Too frequent a schedule of method changes (less than a year is too often for my taste) does increase a perceived uncertainty in the stability of the new application, and also increases the effort in training and communicating with users and stakeholders.

In addition to this, you should design and document your thinking so that your successors can review and understand what you thought when doing this work. The data models are stored in a "Solution Architecture" database in your architecture tool, and the meta model is created in a separate "Config" database. This gives you the opportunity to add future enhancement ideas in a "next state" model after the initial release/during the year that will be ready when discussing changes in the next release with your stakeholders.

This article looks at the first of the three major activities when configuring the tool: creating the data models, designing the meta model, and changing the actual configuration (based on your tool choice, this can be accelerated significantly when you have a meta model in your "Config" database).

Data models as the foundation for the implementation of the architecture tool, as well as interface design

One observation in projects is that clients like to skip over this phase, or do this in a non-graphical way, while showing a mysterious "data dictionary" in a long spreadsheet, that mostly scares away non-technical stakeholders and team members. There is a better way to do this, and having good-looking and easy-to-understand data models is very helpful to bring everyone on the same page (literally).

The first step is to take a look at your **program objectives** and which **architectural views and artifacts** need to be created. Then it needs to be determined which information needs to be captured and how this information is related to each other. Some of this is determined by laws and regulations, other areas are free to be determined by the program. For the latter group, a good start is to look into the chosen tool's method help. All professional architecture tools come with an underlying method that is documented.

The outcome of this is a list of items like "regulatory obligation", "process", "vendor", or "application", and most likely you already thought ahead where this information is captured and how it is broken down. Stop this for a moment, and rather consider this to be abstract items and determine how they are related to each other (and note down your thoughts which app is housing which info on a piece of paper for later). Capture these in business terms, and do not refer to them in the technical terms that an existing tool might use (the mapping between the logical data model and the physical data model, how an application stores data in tables in fields, comes in a future solution design later).

For example: a process is done by roles that follow a series of tasks that are supported by applications and might have risks associated with it. Does that mean that all these things are properties of the "process" entity (a data modeler's slang for a "thing")? No – an application is different from a process and can be used in multiple processes, a role is an organizational element that can be aligned to positions which then roll-up to organizational units, and a step is part of the process on a lower level. Risks can be associated with many of these elements, such as application risks ("what happen if this app fails?"), operational risks ("what happens if this process or step is not executed correctly?") or organizational risk ("how risky is doing a project with this vendor for my business?").

This indicates that it makes sense to distinguish between the "things" and the relations these things have to each other.

Create a high-level data model in your architecture tool

This brings us to the second step of this exercise – open up your modeling tool and create a high-level logical data diagram that shows the relationships between the entities. Do this in a new database ("Solution Database") that you will later enhance with additional models outside the data modeling activity, such as governance processes, or permission matrices.

Figure 18.1 below shows an example of a high-level data model.

HIGH-LEVEL DATA MODEL

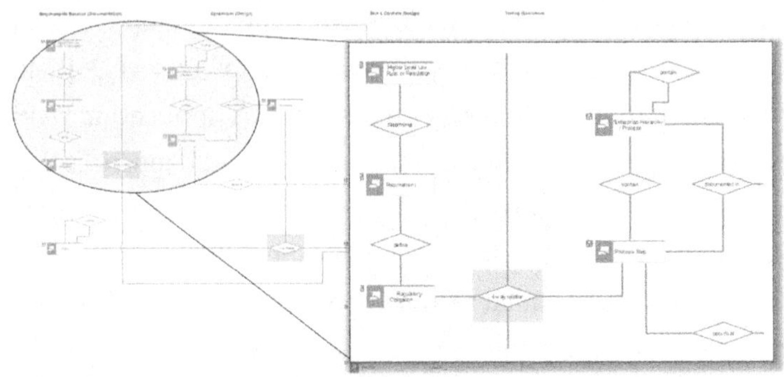

Fig. 18.1: High-level data model capturing the architecture "things" (entities)

When modeling the relationships, think of the following patterns:

- **Direction of the connection.** Is it bidirectional or just one direction (an application impacts a process; does a process impact the application as well?)? Draw an arrow between the two entities.
- **Does your entity stand alone, or can it be decomposed** to more detail on multiple levels (processes or org units typically are decomposed on multiple levels)? Do not draw these multiple levels in the high-level data model, but rather depict this with a self-referencing arrow (see "Enterprise Hierarchy/Process" in figure 18.1 above).
- **Identify the relationship type between the entities.** Is it just a single relationship between two entities, or a complex relationship between multiple objects? The rules for these complex relationships need to be understood and documented so that you can reference them later.
  In addition to this, you can also add the cardinality to the relationship – is it a 1:n or an n:n relationship? An example of an n:n relationship is the relationship between applications and processes; a process can be supported by multiple applications, while an application can support

multiple processes.

- **Can you categorize the entities into "buckets"?** In figure 18.1 above, all regulatory requirements are captured under "Regulatory Baseline" while processes, apps, etc. are captured under "Operations (Design)". This categorization can determine your model layout and makes it easier to understand for non-technical stakeholders.
  Make it easy for your readers to understand how you laid out the entities in the diagram – what is the story you want to tell?

Once you have created your high-level conceptual data model, plan a workshop with your stakeholders to verify and socialize your result, and get their formal approval. Getting the stakeholder buy-in to this high-level model is the most crucial step for the tool configuration and future development efforts. Be prepared that you still will get requests to add stuff once they fully understand what you will configure – the data model might be too abstract for some stakeholders, and they won't admit that at this point.

Add details to the entities to the next level of the data model

The next step is to create more details for each entity:

- **How is this entity described?** Typically, you find technical attributes, such as an ID that a system uses to uniquely identify a record of the entity, and business attributes, such as a description or frequencies (for events), or time attributes (for processes). Capture these in lower-level data models as shown in figure 18.2 below, and add attributes to these that make sense for you — but not necessarily technical field names that a system might provide. Even though, it might be a good start to go through the screens of your application and note down the attributes that are captured there. Just don't stop here in your thinking – if a system does not provide that attribute, then you've identified a future enhancement requirement! You might add a user-defined attribute that suits your needs in the future.

- When you intend to interface with other systems, you might want to take a look at the data model of that system and create a **separate lower-level data model of that system**. Most likely, you will see that these systems capture way more information than you will need for your purposes. However, these external data models will be reused when it comes to the interface design – see below.
- Also look at the **format of the attributes**: is it a regular text field that is needed (and how many characters?), or a number or date field? What is the format of these? Add this to your attribute object information, and your solution architect/developer will thank you later when this attribute needs to be customized in the application.
- A last step here is to **identify the key and foreign key attributes**. What is the unique attribute of your entity that identifies the entity? Are unique IDs of other entities referenced in your entity and therefore need to be identified as foreign key? In figure 18.2 below, you see the majority of attributes shown as normal (descriptive) attributes, while "Item Type" is a key attribute for the process entity, and "ARIS GUID" is a foreign key.

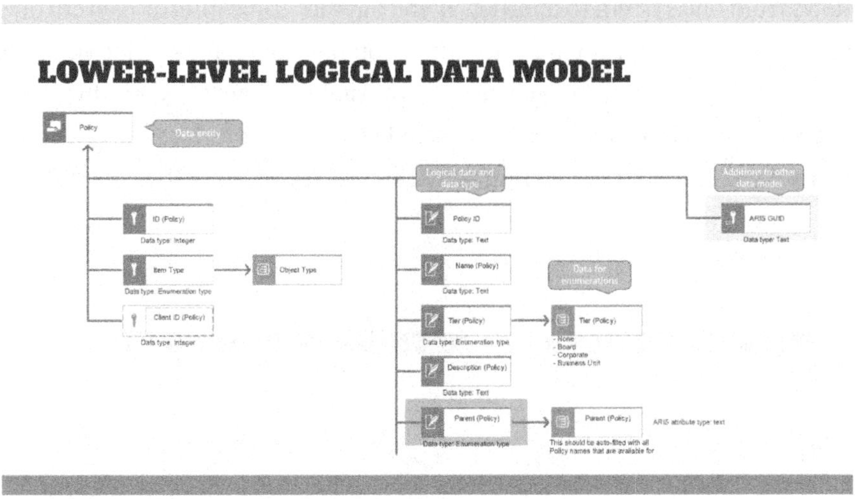

Fig. 18.2: Example of a lower level data model for "Policy"

Once this step has been accomplished, socialize the lower-level models with your stakeholders again and capture their feedback. In this round of review the technical stakeholders, e.g., owners of other applications, have a more prominent role, and most likely want to run these models by their solution architects before providing feedback.

Define the data requirements for technical interfaces

The attributes of the lower-level data models will be used for future mappings in the interface design phase, as well as the identification of tables and fields (physical data modeling) in the future. Figure 18.3 below shows an example of an interface mapping.

You might have to create additional interface models, which might include the mapping of entities to model types. For example, when you have incoming interfaces from other systems of truth ("in which model type will these objects show up?").

In addition to this, you might also want to create technical application collaboration diagrams showing how the interface is set up and which data flows from one system to another. If you have multiple interfaces from one system to another, for example, when you have a data lake that you pull various information from, you also might want to create separate diagrams showing which information will be included in which interface (more details on this and other data modeling activities are described in the appendix).

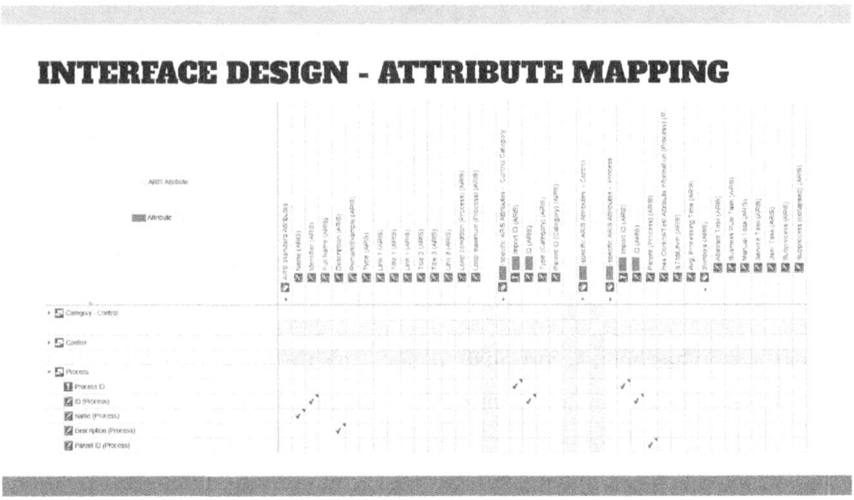

Fig. 18.3: Interface attribute mapping

For documentation and future reference purposes, I recommend starting a "Solution Architecture Overview" presentation that can create links to the data models and future additional models, and update this with each major release. Stakeholders can use this deck later when it comes to detailed workshops or other decision points in the future.

Next step: create a meta model

By now, you should have gotten your arms around what shall be captured and shown in your architecture tool. We will now take a look at how to create a meta model.

What is a meta model?

While the data model shows the "what", a meta model will show the relationships between the models and objects that will be configured in the tool – the "how". It is good practice to create this meta model in a separate database ("Configuration DB").

To create a meta model you need to find an agreement on which architecture views and how many levels you desire to have. An indication of this can be the decomposition that you created in the data modeling exercise, or by looking at the structure of your organization breakdown. Figure 18.4 below shows an example of a client meta model and how a user will be able to navigate from a central start model to a lower-level business process on level 4 that might have subprocesses and its details shown in additional separate models for each step of the process.

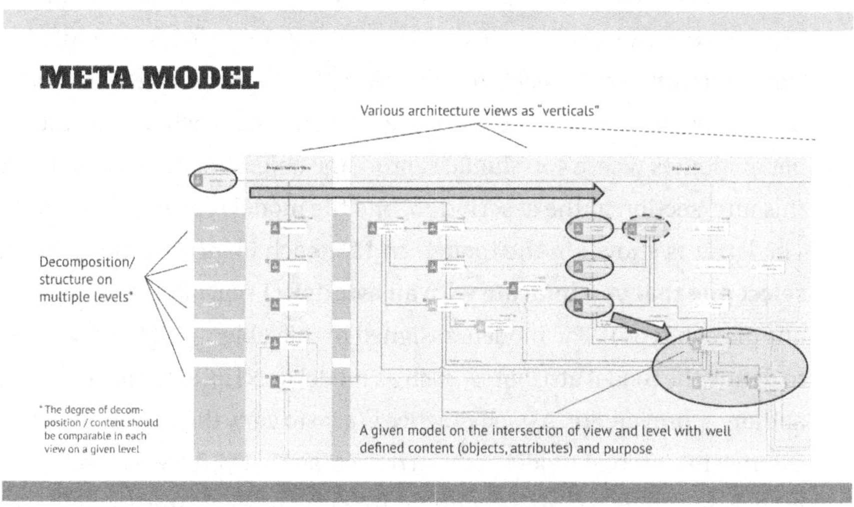

Fig. 18.4: Example meta model

A meta model is the visual representation of "how things are wired up" when the configuration of the architecture tool will be done. This helps to communicate with stakeholders, and other programs that you want to have technical interfaces with. In the latter case, you can derive requirements for levels based on the structure that they might be able to deliver (or require) for data exchange with their tool.

How to create a meta model

In your tool, create a new model that allows you to assign any model type to an object – in ARIS, this model is called "Structuring Model" and the objects in that model type ("Structuring Elements") allow any model assignment. Do this in a new "Configuration" database, which will only contain the meta model and other config-relevant items, but no design or additional documentation. Those should go into the "Solution Architecture" database that also houses the data models.

- In that model, **create a matrix** of views – the light gray verticals in figure 18.4 above – and levels – the darker gray rows in the background.
- Now **place objects on each intersection** that represents a model type that needs to be further defined. Please note, that each of the objects has attributes where you should capture the objectives of this model on this intersection in the description, and the model type in the remarks.
- The latter is shown in the model, and for each identified model type, **select one that you indicate with an asterisk** (I typically take the first one of each type). The models assigned to this object will then be used to define the model attributes, such as model description, model status, author, approver, etc. A good practice is also to copy the objects with the asterisk in a second model as occurrences, so that you have all relevant models for model attribute configuration in one dedicated model.

Once you have the views and levels defined, connect the objects:

- For the **click-through paths** (for example, assignments between higher-level objects and the next-level model) draw a directional solid line. This also shows if a user has to decide when navigating – in the example above, at the level 4 product level, a user can either navigate to the details of a lower-level product (e.g., who is responsible/uses this product), or to the first level of a product-oriented process view.

- To indicate the **reuse of objects** that are created in a specific view (and stored there) draw a dotted directional line. An example can be seen for the lower level process detail model (details of a single task) that has a few incoming lines in figure 18.4 above, but many dotted incoming lines from org charts, application catalogs, or risk and control catalogs. This will keep the architectural setup by view intact, and will allow a detailed permission deployment in the future, e.g., having only users from the IT organization update the application catalog, while business units can create process models that can use applications from the catalog.

Overall, a meta model is very helpful in communicating "how things are wired up" and will be the foundation for future scripting work, as well as the configuration (see below).

Create the next level of models

Once you have a draft meta model, the next step is to create a level 2 model assigned to each of the objects in the meta model. Within this model, all objects, and connections will be modeled using generic objects, such as "task", "event" or "application". This will prevent stakeholders from looking at the content instead of the structure of a model.

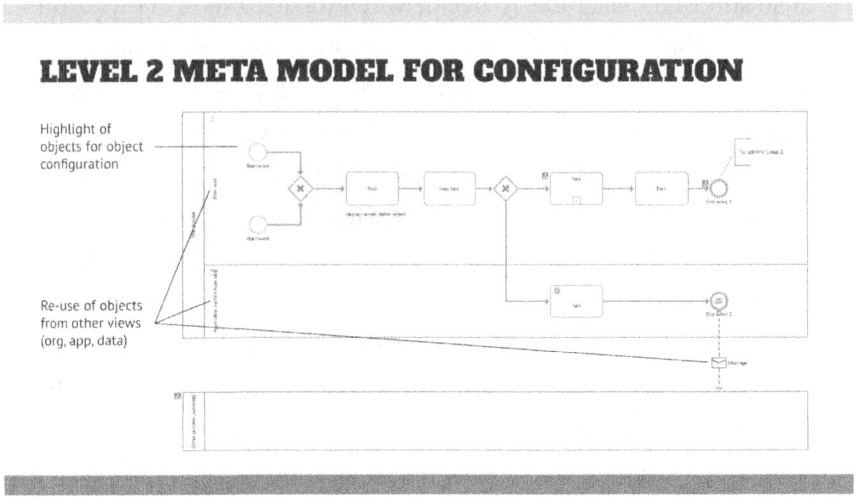

Fig. 18.5: Level 2 meta model for configuration

Some guidance for modeling the level 2 meta models:

- **Use generic objects**. At this time, you are looking for an occurrence of an object or connection in a model. It doesn't matter what the content of this object is. Quite the opposite – if you have "real" content in your lower-level meta models, your stakeholders immediately will look at the content and tell you that you got the process steps or a technical application interface wrong. This means they lost focus (concentrating on objects and connections shall be configured later).
- Create connections and s**elect the correct connection type if multiple connection types might be pre-configured** in the tool's Entire Method. You might have to **create additional model snippets below the main model if you want to include additional connection types** – for example, when you want to create the four RACI connection types (responsible, accountable, consulted, informed). The tools typically allow only one type of connection between two objects, so you need to create additional modeling snippets for the other connections. It does not matter if the model is not "pretty" after that, but it is important that the connection

is defined in the database when it comes to creating a configuration file later.

- **Reuse objects from other views**, such as apps or roles, in your model to create a complete picture – but change those only in their correct view. You want to avoid having multiple objects of the same kind being configured differently, which will make the debugging of your configuration files significantly easier.
- **Highlight objects that need to be configured**. I typically use the first occurrence of an object type in its native view and put a light gray box behind it. This is the visual marker for later when the configuration shall be changed so that you don't have to hunt for the correct object. Unfortunately, not all tools allow copying any object type into a "generic" model type, so you cannot create the convenient separate model that you created for the model attribute maintenance above.
- **Maintain object attributes in the highlighted objects (and only there!)**. This includes attributes like "description", or "remarks" but also specific attributes that you will need to create for interfaces or are used in special use cases like simulation. You do not have to put meaningful content into these attributes, a simple "x" will do the trick – you must have a maintained attribute in your config database at this time.
- **Create assignments** from the objects in your lower-level meta models to other models according to the high-level meta model, so that you can navigate vertically and horizontally in your config database. Furthermore, please ensure that you use occurrences of events, for example, when you "wire up" the horizontal navigation of processes at level 4 for example.

These generic lower level meta models then can be used to discuss with stakeholders. During these conversations also identify their requirements for formatting (e.g., colors), and also if attributes shall be shown in the model (for example an indicator that shows up under certain conditions – i.e., an indicator that shows if a risk or control is related to a process step in a lower-level process). This information will be used later when creating a template in which you define the look and feel of your models.

Create examples to make things "real" for your stakeholders

After you have discussed the "generic" meta models with your stakeholders, create models with "real" content that can be used in training or communication material later. Let the stakeholders point you to the subject-matter experts for the different architecture views, and work with them to create meaningful content.

In an ideal scenario, you can develop a story how you would walk new users, or representatives of other programs that you want to interface with, through the meta model that will make sense for them. Pick a generic example that everyone can relate to – "onboarding" or "purchasing" processes are well known by almost any user.

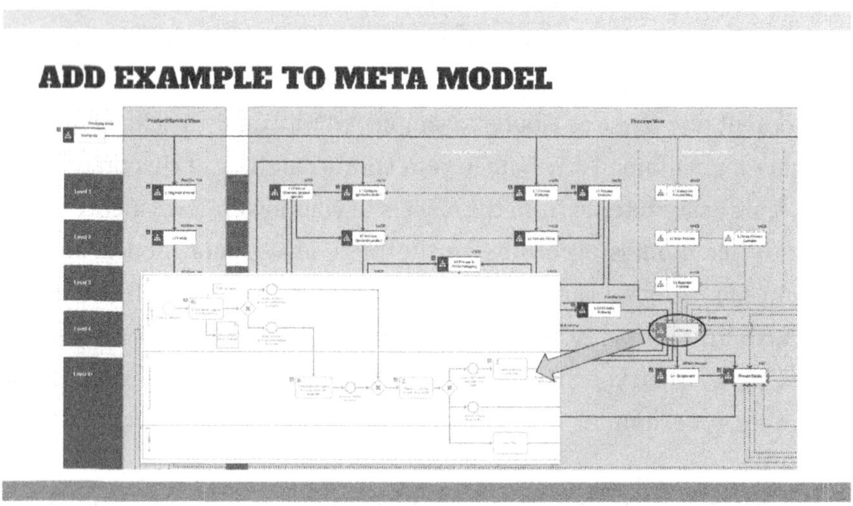

Fig. 18.6: Add example diagrams to the meta model

A leading practice is to have both assignments on the objects in your meta model – the "generic" models that include all objects, connections, attributes, etc. (for which I like to add a "- MM" at the end of the model name), and the "real" examples, for which I typically add an "- Ex" to the model name.

Once you have these two main steps –the meta model itself and the assigned model types – accomplished, review the results with your stakeholders and gather feedback. This will create a shared understanding, uncover the need for a method extension in the form of new symbols or attributes, and point out potential mismatches between tools (which will become requirements for future releases), as well as requirements for the template creation at a later point.

Similar to the Solution Architecture documentation, I also recommend creating a "Configuration" presentation that walks stakeholders through the major areas of the meta model and includes links to the models in the Configuration DB.

Create the configuration files for your tool

The last step in this series is to use the information above and configure this in your architecture tool.

Creating the architecture method configuration

After doing the "hard work" of figuring out what shall be captured and visualized in your architecture tool – manifested in the data model and meta model – it is now time to do the actual architecture method configuration of the tool. For this purpose, I will describe this for ARIS. Your chosen tool might differ in the detailed steps, even though the major steps most likely will be the same.

Moreover, please do not see this as a detailed "where to click" guidance. For that, I recommend that you attend a training either by the vendor or your consultants of choice. This chapter will give you the overview of what to look for and leading practices when working with the ARIS method configuration.

Architecture method configuration process

There are three different types of "products" you will have to manage in ARIS (your configuration files): the method configuration file (the "**filter**"), the **template** that defines the look-and-feel of the models, and the **configuration set** of the portal (ARIS Connect).

We will discuss the approach below, but keep in mind that you should treat these artifacts like you treat customer products – they have their lifecycle (i.e., version them), and you should never make configuration changes in your production environment. Rather have a setup that includes at minimum a development environment, a test environment -which you can use for training purposes as well- and your production environment. Please note, that the lower environments might or might not have the same (logical) database structure as the production environment.

The process for architecture method configuration is as follows:

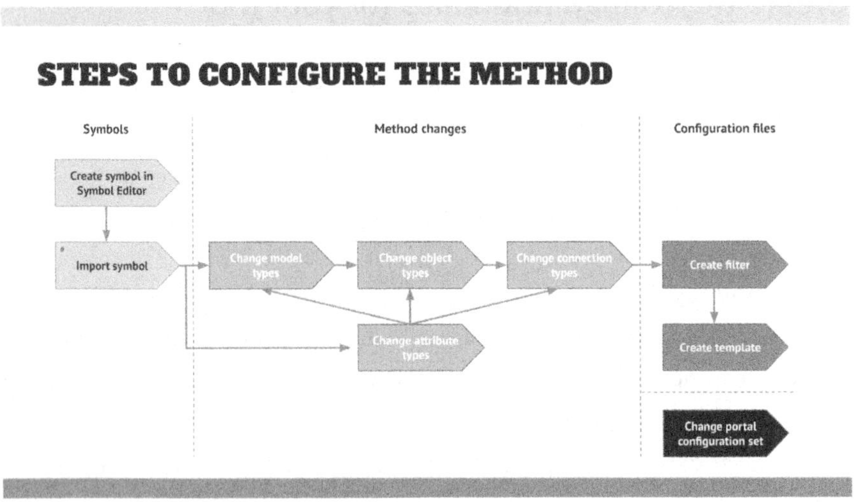

Fig. 18.7: Method configuration approach

This process is accompanied by additional activities such as checking and updating of scripts (reports and semantic checks), documentation, and packaging for deployment.

For all the steps except the last one (Change portal configuration set) you will need the ARIS Architect download client at the time of this writing, even though parts of the method configuration are also available in ARIS Connect's administration, and I expect to have everything in ARIS Connect within the next few releases. The portal configuration will be done completely on the ARIS Connect website.

Create and import symbols

ARIS ships with many symbols, and you can create your own symbols or change existing ones in ARIS Symbol Editor – you will find the symbol editor in the "ARIS" menu of Architect. Adding a new or changed symbol to your configuration consists of two steps – create the symbol and import the symbol.

Create/change symbol

Within the symbol editor you will find all the tools to create or change a symbol, and you can get started right away if you have worked with similar tools before (PowerPoint for example).

However, one thing you need to understand is that the default symbols are a complex set of shapes, and you need to be careful where to click to select the right part of the symbol. Figure 18.8 below shows how a symbol is constructed – it is all vector shapes, except the icon that is in the colored background vector – that is a bitmap graphic (the small flag in the graphic). If you want to change that, I recommend creating a bitmap and not trying to do this within ARIS. Most likely, you will run into scaling effects that make that small icon look bad (guess how I know...).

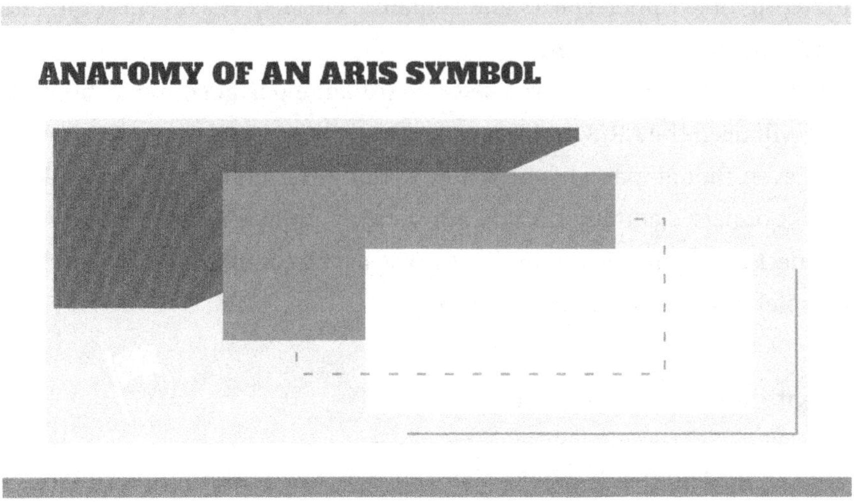

Fig. 18.8: Anatomy of an ARIS symbol

When you have created your own symbol, play with the zoom and check if your new symbol looks good at 100% -75% and not only in the zoomed-in standard display of the symbol editor.

The second step is to create the small symbol for the display in the symbol toolbar and in the Explorer view. You can find this on the second tab of your symbol in the symbol editor, and it typically used the small icon that is used in the main symbol (the flag in the example above).

Once you're done, save your symbol as an *.amf file in a location where you can find it quickly again for the next step. I typically use a descriptive name, such as "flag orange" in our example. When you create symbols for attributes, you might have to create variations of the same symbol, so you might also have a "flag blue" and a "flag gray" *.amf file to create to display the three states of a Boolean attribute (true, false, not maintained).

Import the symbol into your configuration

This step is pretty straightforward – you need to import your *.amf file into ARIS.[46] You can do this under Administration > Configuration > Method. The one thing you need to decide is if your symbol is to be used as an object symbol (the large boxes in your model) or as an attribute symbol (small icons inside or around your main object symbol). Keep in mind that your attribute symbols are significantly smaller than the object symbols, and you might have to go back to the symbol editor to change your *.amf files.

To import your symbol as an object symbol, just go to "Symbols" in your method area and right-click on the right-hand pane and choose "import". In the next dialog, you can select your *.amf file. At this time, give your symbol a "meaningful" name, such as "Business Capability Configuration" because that is the name that you will see for the object in the toolbar.

If you want to import your symbol as an attribute symbol (a different section in the "Method" area), I recommend looking through your list of already existing attribute symbols and derive a new one from it (right-click on the existing attribute symbol). In the following dialog, you can upload and name your new symbol. Typically, I like to derive a new attribute symbol from the "Check Box" attribute symbol. Like with the object symbols, give your new symbol a "meaningful" name.

Changing the architecture method

The steps in this category are the actual changes in the configuration and encompass the change and creation of new "things" for attribute types, model types, object types, and connection types. You will also prepare the Configuration database to be able to cut a filter in the final high-level step.

From here on, you can do the configuration in either Architect or directly in Connect. However, there are still some things not included in the Connect method configuration (as of the date of this writing), and therefore I will describe things as if you would use Architect.

Create attribute types

This is one of the more tedious aspects of the method configuration, and your lower-level data models will help you do this in a way that you don't forget things:

- Create **attribute type groups** – even if those are not visible in Connect, the attributes will be sorted out in those groups (and are visible in Architect). This will help you to organize the attributes and avoid a super-long list of attributes when it comes to maintaining them. This is optional and recommended, but you can search for attributes in both tools in case you encounter a long list.
- Having all custom attributes from an external system, for example, in a separate attribute group will also help with the negotiations and the updates of the data models going forward.
- Create **new attributes/update existing attributes** – when you create a new attribute keep in mind that you cannot change the attribute type (text, integer, boolean, value, etc.) after you created it. So you will have to recreate the attribute and then assign it to models, objects, and connections again (see the next steps below). Therefore, it is important that you discuss that format in the data model conversations early on. When you choose "Boolean" as attribute type you can not only define the values – true, false, not maintained – but you can also define which of the freshly imported attribute symbols shall represent the value if wanted (the actual definition if it shall show up as text or as the symbol will happen when editing the template in the wizard).

Figure 18.9 below shows an example of the data model, and you can easily see the custom attributes that needed to be created in this example for a technical interface (the 11 attributes to the right).

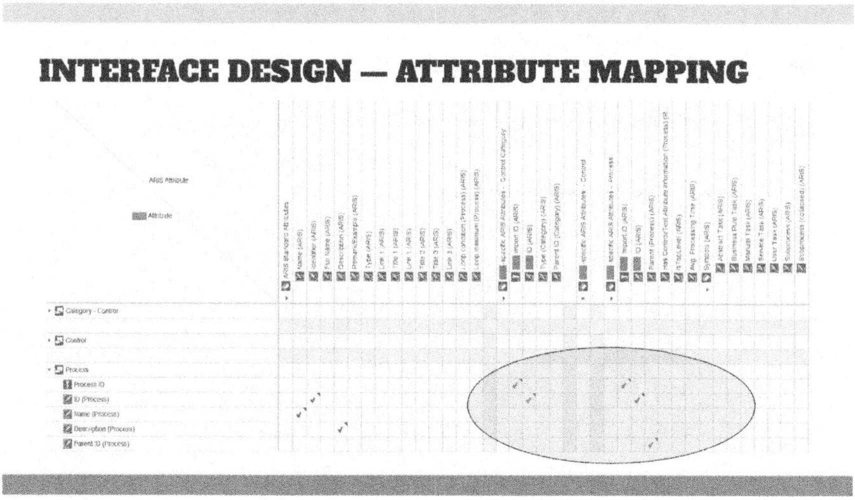

Fig. 18.9: Identification of to-be-created new attribute types in the configuration

After you have changed the architecture method, you want to check if they show up as an option in your "MM" models of your meta model. Remember that you have to restart your Architect because the method gets loaded when the tool starts. In Connect, a restart is not needed.

Change the model types, object types, and connection types

The next step is to go into the other sections of the architecture method in the ARIS Administration and make the changes that you've determined in the data and meta modeling activities:

- Change the model types and derive new ones if needed, for example if you want to have different levels of Value-Chain Diagrams in your method so that you can specify queries more precisely.
- Change the object types as needed.
- Change the connection types as needed (renaming only as of the writing of this chapter).

METHOD CONFIG IN ARCHITECT AND CONNECT

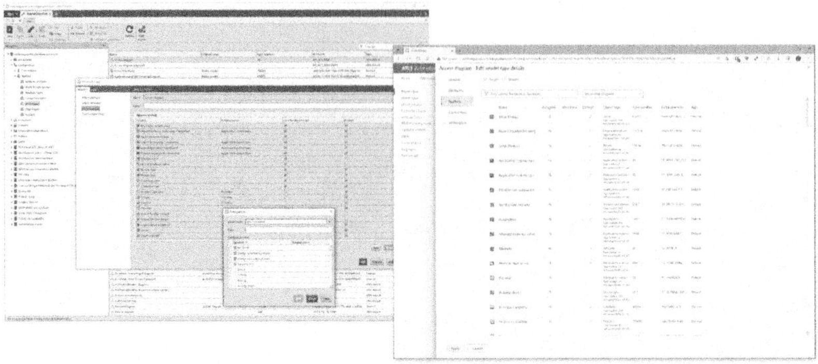

Fig. 18.10: Method configuration in ARIS Architect and Connect

Each dialog that you will see will have different options that vary depending on what type you edit. They all share the first page of the wizard and the attributes page, though. Keep in mind when you do this that there is a hierarchy, and you make all changes in the context of a model (as the highest "method thing" in ARIS), which means that you can define a connection between two objects in one model type, while you do not allow a connection between the same objects in another model type for example.

The next step is to go to your Configuration database and maintain all model, object, and connection attributes in the objects that you've highlighted in your "MM" models, as well as the model attributes in the models that you created the Model Overview model for. Go back to the meta model section in this chapter if you might have missed these two recommendations.

You don't have to put any meaningful information into the attributes, a simple "x" is sufficient. You must have an entry in the underlying database, and it doesn't matter what the content is. Furthermore, click all Boolean attributes, select a value from value attributes, enter time and currency attributes and so on, until you have all attributes that you've identified in your data models being maintained.

Update the configuration files

The first step in this section is to create a filter that includes the changes in the method. Subsequently, you will determine the look and feel of your models by editing the template. Lastly, the configuration set of the ARIS portal needs to be configured.

Create a filter

Now that you have made all the method changes in the backend, it is time to create the filter. There are three ways of creating a filter:

- Use the **filter wizard**, which is a lengthy process that gives you control over every aspect of what should be included in your filter.
- Create a **filter from a database** – this is the recommended way of creating filters, and the reason I recommend the creation of a Configuration database.
- **Merge of two filters** to create a third one.

Now go and create a filter from the Configuration database (ARIS > Administration > Configuration > Filter). Give it a name and add a version number to the name. Use the description field to add a version history – I typically keep it simple and follow the standard "YYYY-MM-DD: [insert change descriptions or 'initial filter creation']".

Then run the filter report (see below) on your new filter and see if all architecture method changes are included in the filter. If not, go back to the configuration database, or even back to the Method section in Administration to make the necessary changes. Update your filter and run the filter report again. This can be tedious work, but the fact that you can use Excel functionality like counting rows is very helpful (for example, when you want to check if all models have the same number of model attributes).

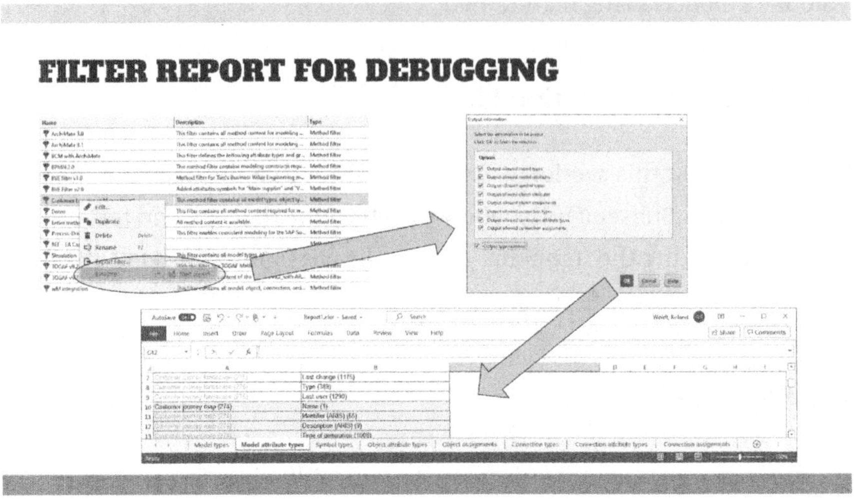

Fig. 18.11: The filter report helps debugging your configuration

Merge filters

In some scenarios, it might not be sufficient to have a simple filter creation from the Config database. If you work with complex notations, such as BPMN, it might be a good idea to not include that notation in the "raw" filter, but rather copy the specialty filter that ARIS ships separately, and make changes in the copy using the filter wizard. Then merge both filters into a third filter.

One can complicate this as needed, and figure 18.12 below shows the scenario that there is a default filter defined for the whole organization, while a project has additional needs (for example to include model types that are only of interest to them). In that case, you would create an "Extension" database similar to your Configuration DB, cut a filter from that database, and merge it with the Standard filter into a Project filter.

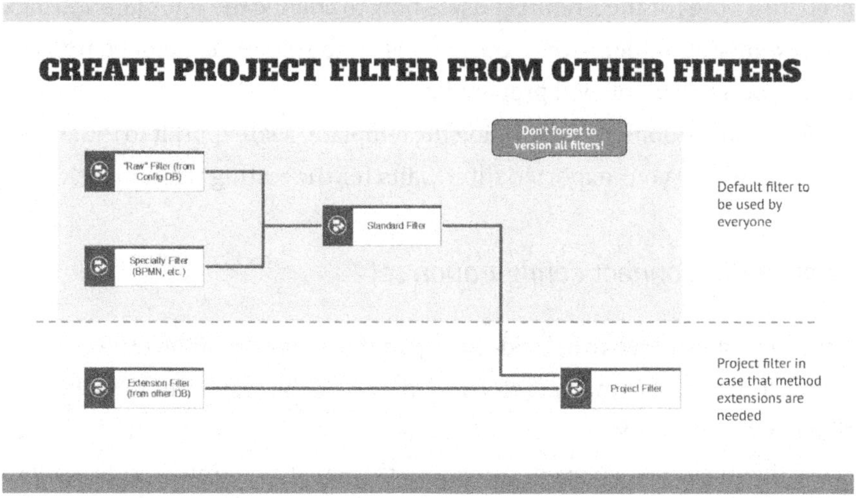

Fig. 18.12: Creating a project filter from merging different filters

Once you're done with the filter creation, export all to a shared location, from which the testing team can pick the latest version for testing purposes (or you package it for them).

Change the template

The next step in the architecture method configuration is to change the look-and-feel. This is done by using a template in ARIS, and the steps are supported by a wizard that walks you through the various aspects of the visualization changes. Please note, that in the first step you need to select your final filter – the Standard and Project filters in figure 18.2 above – as the method filter, so that you just see the objects and connections that are relevant. By default, this is set to "Entire Method" whenever you call up the template wizard.

Also maintain a version number and version history on the first wizard screen, just as you did with the filter(s).

Once you have completed the template definition, you need to set the default template for your Connect users in Connect administration, and prepare a note for the Architect users how to change the template settings for new models under ARIS > Options. This note shall be included into the release notes that you will prepare later.

When you are done with changing the template, also export it to the shared location, where your exported filter waits for the testing team to pick it up.

Change the Connect configuration set

The last changes need to be made in the configuration set of the ARIS Connect portal. There are too many options to cover here, but you want to make sure that any renaming of model or object types will also be reflected in the config set so that they are displayed correctly in the screens. You also might want to check if your changes – in the case of method extensions – require changes in the fact sheets and search user interface.

Unfortunately, the filter and the portal configuration are not connected because the target groups are different (viewers and standard modelers in Connect, and power user and admins in Architect), and the idea is to not confuse the viewers/standard modelers with too much jargon.

Export the configuration set to the shared location from above. When you're done, you have completed the technical architecture method configuration.

Final activities to close out the architecture method configuration

You are almost done. The steps below technically don't belong to the architecture method configuration, but are "good development behavior" and you should follow them as well.

Update scripts

The last thing that you want to determine (and change if needed) are **reports and semantic checks**. The latter are meant to support users in correct modeling, and your method changes might trigger "false negatives" if your scripts have not been updated. If you need to update those, don't forget to also **update the script documentation** and prepare a **guidance document on how to replace the old scripts with new ones** (**and when to schedule them**) for your IT administrators.

Additional documentation and activities

At this point, it is time to document the changes that you did. Typically, this will be done by creating a **"Methods & Conventions" (M&C) document**. This can be done with a regular Word or Google document, but I highly recommend having this available in a Wiki, such as Notion,[47] so that you can update and publish new versions of the documentation quickly.

In addition to the M&C document, you also want to create **release notes** that complete the product package and can be used by the org change management team for communication purposes.

Test and deployment of configuration files

If your organization follows a more traditional, formal release process, you also might want to **package all configuration files, scripts, and documents in a zip file** for sending it over to the test team and IT administrators.

The configuration files then need to go through testing in your test environment before they will be released into the production environment. Be prepared to give your testers and technical personnel some 'hand-holding' attention when they need to create their test scripts ("what is the expected behavior?"), conduct testing, and deploy into different environments.

19

Architecture Database Setup And Structure

Setting up your architecture database structure and the internal structure of an individual database is a task that is done very early in an implementation (and you are using a database-driven architecture tool, aren't you?). It is defined in the "Technical Governance" phase, and typically is considered an afterthought, even though it has a large impact on how easy your users will perceive and use the tool.

Let's take a closer look at what needs to be thought of when designing your database structure and overall setup.

Use cases

The first thing you should think about are the objectives and processes that you want to support with your architecture database setup (and to be clear, we are talking about logical databases, and not the physical database where your tool stores all information, which is typically just one per instance). Some of the most common use cases are:

- Release management
- Solution architecture and design

- Import of existing content
- Provisioning of Reference Content
- Archiving models for later use
- Synchronization with implementation and runtime systems

Typical architecture database setup

Even though you could set up your architecture program with just one database -and the newer tools support this, and got rid of separate publishing servers, it is recommended to have a multi-DB setup. This allows you to set up a proper release cycle, and automate all activities in that process as well.

Figure 19.1 below shows the minimum recommended setup in the gray box:

- A **"Work DB"** as the only place where changes are made. The expectation should be that any content might be different tomorrow, and that only admins and modelers have access to this DB. Once a modeler submits a model for approval, it will be moved into a Review DB.
- The **"Review DB"** is a temporary location that is only used for the single purpose of being the staging area for reviews and preparation for publishing (following the release plan). The users of this DB are the stakeholders who play a role in the approval process. These stakeholders can be business users for content approval, SOX reviewers, and technical QA folks who make sure that all models adhere to the formal standards of the organization. This is so that all models allow for proper analysis and also have a consistent look-and-feel.
  Depending on the approval or rejection, a submitted model will be either sent back into the Work DB for changes, or will be versioned and merged into the Master DB. After the approval of all models in the Review DB, they can be deleted so that the database is pristine again for the next release of models.
- The **"Master DB"** is the place for all non-modeling users and consumers,

who most likely will just need a "view" license. The content here is the "official architecture" of the organization in the form of an up-to-date Enterprise Baseline (see below), and can be used for reporting, training, regulatory and other purposes.

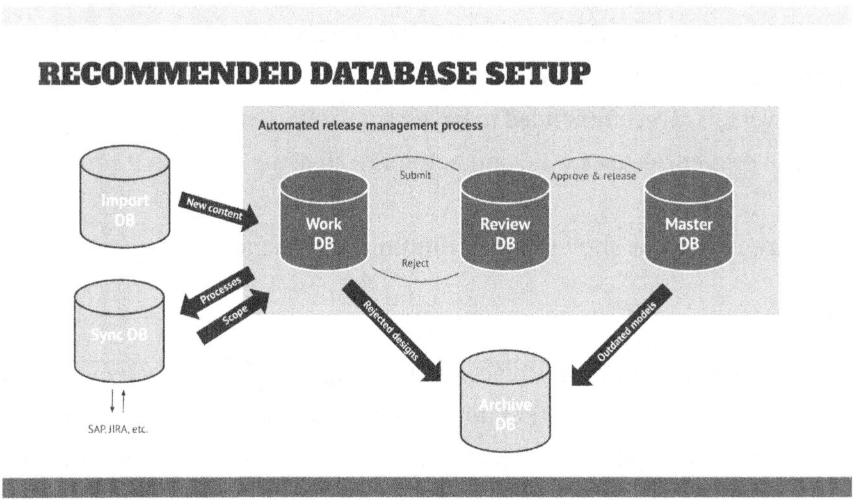

Fig. 19.1: Recommended database setup

Besides this core DB setup, I recommend having additional databases as needed (green in figure 19.1 above):

- An **"Import DB"** – when importing content into your architecture tool, it expects an identifier that it uses to manage all models, objects, and connections in the database. For obvious reasons, existing content that you want to import from legacy files (Visio), or other systems (e.g., any other tool that exports *.bpmn files) don't have this information, and also might import additional content that is not needed or contradictory to your organization's standard.
  This database will be used for the "raw" import of models, and some

initial clean-up before only the desired content will be merged into the Work DB for further normalization and integration.

And just to be clear – if you export content from your Work DB, for example for translation, and want to re-import it into your workspace, then you don't have to use the Import DB. These export/import reports maintain the individual IDs. Therefore, the tool recognizes the content during import and updates it accordingly. This is the preferred way when creating interfaces between the various systems of record and your architecture tool.

- A **"Synchronization DB"** – some architecture tools allow a bidirectional synchronization with implementation or runtime systems (e.g., SAP's Solution Manager). The risk of a bidirectional sync directly with the Work DB is that someone could have changed the scope in the implementation tool. This would trigger the deletion of an object definition in the Work DB, while that object might be used in other models as well. For obvious reasons, this is an undesired behavior – it will create "holes" in other models that cannot be easily restored. The use of a Sync DB allows you to do a sync and compare the result with the content in your Work DB before you merge the sync result into the Work DB.
- An **"Archive DB"**, which will hold content that is not relevant anymore, but should be kept for future purposes. Example content can be projects that never made it to completion, but might include things that might be picked up at a later point. For example, when the technology that was researched in a project has matured so that it is feasible for use in an enterprise.

  The main benefit of merging the content out of the Work and Master DBs is that reporting and analysis will not show false results due to the outdated content.

  Please note that you do not have to archive models when you use the versioning functionality of your tool. This will make sure that a copy/previous version will be saved, and the content will be overwritten as part of the release cycle process mentioned above. This allows you to compare your latest model with earlier versions in your tool and is preferred over manually moving models, even if your tool still allows you to compare models in different databases.[48]

- The last database(s) that I recommend are **Reference Content DBs**. Those are the content starting points that you have identified as accelerators for your projects, but might not have a place in your architecture yet (see the bullet "reference content" below).

If you are developing scripts, dashboards, etc., or make method changes, you might want to do those in separate environments and leave this setup for the Production environment. I recommend having at least two non-prod environments: development and test. The latter then can also — like in almost all IT projects I worked in — be used as the training environment for new architects.

Internal database structure

As mentioned in the "Frameworks" chapter, you might choose to follow a framework, method, or reference content. Typically, these can be used to structure your databases consistently – some of them, such as the SCOR reference model,[49] deliver content in a certain structure that you might want to follow. Others, such as APQC's Process Classification Framework,[50] bring not only a structure, but also related KPI definitions and provide benchmarks (for a fee).

Either way, if you start from scratch, or adapt the content structure of existing reference content, the first thing you might want to take a look at is what your overall database structure should be. Typically, reference content just gives you a certain slice of an architecture, such as processes, but not a full architecture structure.

Once you have defined what you want to do with your database, the high-level folder structure is determined. Typically, you will find two main folders in it: **Enterprise Baseline** and **Projects/Solutions**. The former reflects the "what runs" in all architecture views, while the latter is the place to design the "what changes" (also in all architecture views). Just because of this, it is beneficial to define the structure of the architecture views next, so that this sub-structure is identical in both main areas. That allows you to move content from the "Projects/Solutions" area into the "Enterprise Baseline" area when a project goes live, and it saves you the "translation" of the content to match the receiving structure.

My recommendation is to look for industry standards to determine this structure – a viable candidate for this might be TOGAF's Architecture Development Method,[51] or the ARIS method,[52] which was developed before the tool was created. Adapt the structure of the views and stick with them going forward – your users and viewers will learn this quickly and expect it when they look around in the database(s). So, new learning is not required just because a single project has a different preference.

In addition to this, new projects will be accelerated because they can copy content from Baseline and change it in the Projects area. It will also improve the overall data quality because the Baseline content is approved, and the project team does not start from scratch (and might forget to identify/capture things).

Figure 19.2 below shows the application of this in a project example.

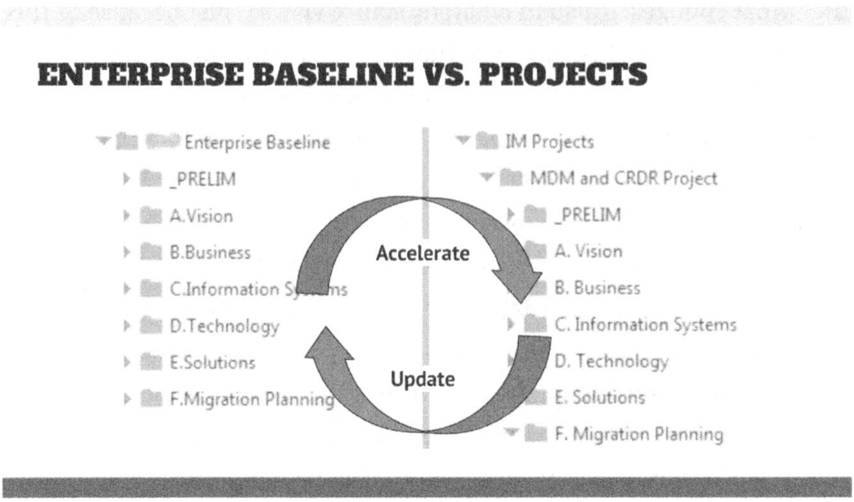

Fig. 19.2: Internal database structure

Once you have set up this main folder structure, you might want to create some more high-level folders:

- **"Reference Content"** as read-only in case you want your users to actively reuse content. This might sound like a good idea at first, but it might lead to the copying of other content that might not be a 100% fit, which will clutter your project's area, and might influence the analysis results, and the risk of using that content when modeling, even though it is not "governed."
  You might want to consider having reference content being provided in a separate read-only database. Have your project teams work with an admin or power user to clearly define what you need, and have them merge it into your project space.
- **"Import"** – this is a temporary folder that is used when importing existing content from legacy formats, that need to be sorted out as part of the normalization activities after the "raw" import.
- I do not recommend creating an ***"Archive"*** folder, but rather use an external database for this purpose. This will improve the analysis results

that you will do in case the user started the report/query in the "wrong" folder, as well as avoid confusion in modeling, when outdated content is always available.

In summary

Setting up your architecture databases and defining the internal structure has its benefits and gives your users guidance and stable expectations of what to find where once they've learned how you have set up your databases. This increases the adoption of your designs, and also enables a common perspective on how things are seen and measured. A release cycle process that includes the initial creation and update of existing models, as well as regular certifications, will keep your content fresh for your users.

20

Library Import And Management

One thing you will notice pretty quickly when collecting content for your architecture efforts is that there is a lot of information in the systems of an organization. Unfortunately, it is not always clear what the "health" of that information is (Is it outdated? Is it correct? Etc.) and if that source that you looked at is the correct one. Or if it is just a collection of items that a project assembled for a very specific purpose. What you found were the "leftovers".

In addition to this, the question comes up, for which data and purpose the architecture repository will be the master. But to answer that question, let's take a look at the type of data that you might need.

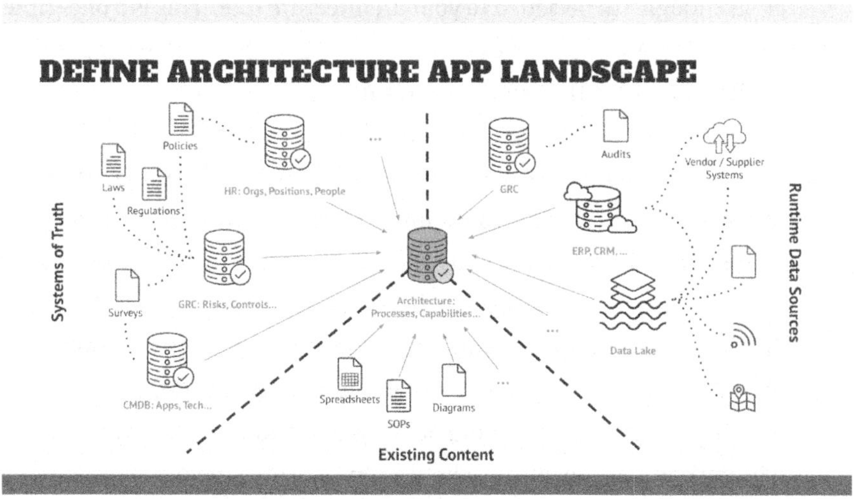

Fig. 20.1: Example application landscape

1. The first type of data that you might need is relatively static data — applications, risks, roles, etc. Those have attributes and might have additional relationships to other objects on top of it, or will be fed by other systems of one-off items, like annual surveys to update the information.
   But the main characteristic is that this data is typically hosted in some form of system/database. You now will have to figure out if this is in one system or many, and which one of those will be the "**System of Truth**".
   You might want to build interfaces to these systems — either one- or bidirectional, scheduled or in real-time — and they will provide you with always up-to-date catalog objects that you need to create your architecture catalogs and drawings. But keep in mind that all changes to that information has to flow from the System of Truth to any other downstream repository ... and your architecture application can be a System of Truth as well.
2. The second type of data is **existing content**, that you determine to be

"good enough" to be added to your architecture tool. This is a one-time upload of content in various formats ... Office files, diagrams, other architecture tools that shall be retired, etc.
This is the key point here: the information will be brought into your system and then discarded. Do not allow this to be updated going forward, since the sources do not include any identifier from your system. Any future update will most likely result in the creation of new objects instead of updating an existing record.
3. The last class of data that you might be interested in is **runtime data**, that you will use for your analysis. This can be execution data to be used for Process Mining, or it could be the latest audit result that you want to import and, for example, show as an indicator in your diagrams.
The main point here is that this is typically considered "disposable data" that you use for a certain purpose, like a Process Mining analysis. This means that you might build data feeds to your stack or work with exports in a file format (CSV, XLS, or similar), but you will not bring in that data as library objects into your repository.

For the rest of this chapter, we will be looking at the process of creating interfaces for Systems of Truth. In the "Real-World Application" segment of this book, you will find the process for the one-time import of existing content. In a future book, I will cover the import of runtime data for Process Mining... a topic that is too big to include in this book.[53]

The process of creating and managing interfaces

The process is split into four phases and one infrastructure topic. Figure 20.2 below shows them in an overview.

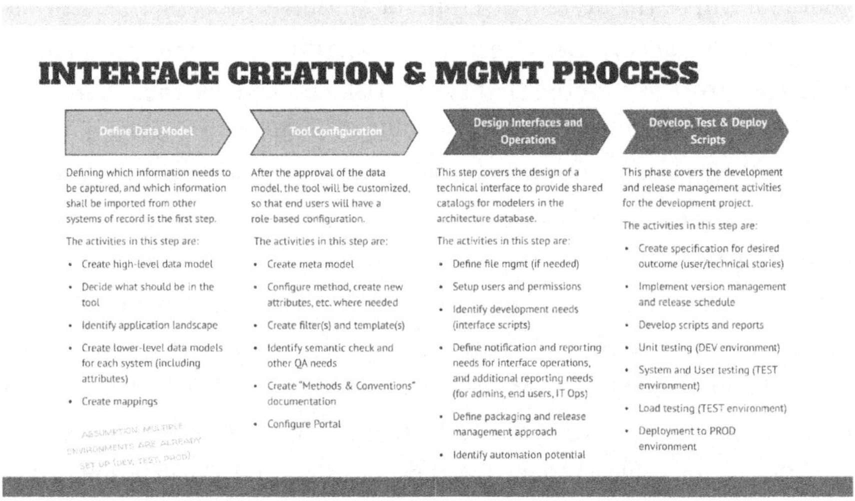

Fig. 20.2: Interface creation and management process

Define the data model

The process starts with the **definition of the data model**. What is the information that you want to put into your architecture tool? We discussed this already in the "Technical Governance" chapter above, and in the appendix you will find more information about how to model data. The key here is to define the high-level entities first to be able to **define the application landscape**. This is what we discussed in this chapter above for the "System of Truth" data.

It is crucial that you have a good understanding of what information you want, but you also need to have a good understanding of your stakeholder landscape, that we discussed in the Strategy section above. This way, you not only understand whom to approach, but also know what "is in it for them" when you approach them with the idea for building an interface to your architecture tool. For a lot of these stakeholders, for example the owner of the Risk & Compliance tool (GRC), there might not be anything interesting to offer from their perspective. However, once you clarify that your tool will be the System of Truth for all things process, that attitude might change quickly. Because now you take off work from their plate to define the processes for the First Line of Defense, and you can assure, that only the sanctioned risk and controls will be available for modelers.

There is no need to worry that modelers *invent* their "own" objects that he might not be aware of (which would endanger his efforts). A win/win for both in my book.

Then you can continue to create the lower-level data models and interface matrices that are described in the "Technical Governance" chapter. This way, it is clear what type of information you will be able to get and which transformations you might have to do. At this point, the technical format of the information (physical data modeling in the appendix) will be important ... especially if your organization has multiple sources for the same type of information and is not willing to consolidate that into one System of Truth.

Set up the environments

Until now, we have just looked at one environment for your architecture tool — the production environment in which you will do all of your work, and to which you will build the interfaces.

However, when building the interfaces, you should follow IT development best practices and have at least two non-production environments available:

- **Development**, which is the environment where you will develop the interface and also all additional scripts, reports, etc. that I mentioned in other parts of this book.
- A **Test** environment, which should mimic the production environment from a configuration perspective so that you can find potential bugs that are caused by different versions of tools in that environment.
  The test environment can also be used to test the upgrade of your architecture tool (in case of on-premise hosting), as well act as a training environment. From what I have seen in the past, it is a "disposable" environment that can — and should — be reset when a new test or other

activity is scheduled.

Figure 20.3 below shows the three environments in an overview.

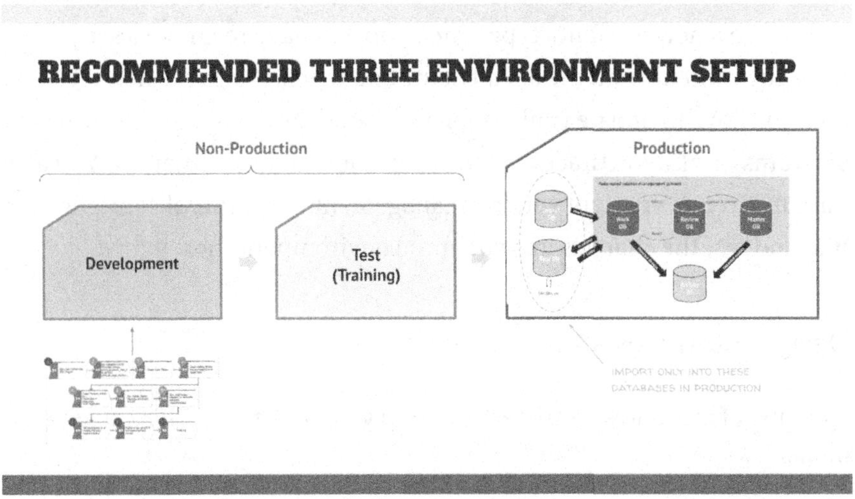

Fig. 20.3: Recommended Prod/Non-Prod environment setup

While developing the imports, interfaces, scripts, etc. always keep in mind that everything is a one-way street. Any new version always goes from left to right and never in the other direction. This is one thing that business people sometimes have a hard time to understand in practice (they all agree to this in theory) and then they are asking for "just a small configuration change" in Prod, that will never be done in the lower environments.

Tool configuration

It goes without saying, but you must have your tool properly configured to store the information that will be provided by other systems. In the "How to set up technical governance" chapter, I described the process for this, so I won't go into depth here. But there is one thing you definitely would like to take a look at — the "physical" properties of objects: do you have the right attribute type? Is the field long enough?, etc.

The challenge with that is that some tools do not allow you to change these definitions "on the fly" when you already have some content created in these attributes. This means that you will have to go back into the configuration and create a new attribute type. Then you will have to create a script that reads the content from the old attribute and writes it into the new one. And then you have to update your template to show the attribute values in your diagrams or other artifacts as indicator... not an ideal situation. You also want to have a very tight version management on your tool configuration files and test the changes in a non-prod environment thoroughly.

Design interfaces and operations

Speaking of attributes: you need to design your interfaces. This covers a few items:

- The **data modeling** described in the "Technical Governance" chapter.
- What will be the **exchange format**? Will you have a file-based interface, aka the Systems of Truth export a file regularly, or do you connect directly to the system or a data lake?
- **How often** do you update? Even though real-time might be tempting, I do not necessarily recommend this. Since there are many people involved in architecture and analyses might take some time, a lower frequency like "once a week over the weekend" might just be fine.
  This also allows for checking the import results and if you have some undesired experiences and want to roll back to the previous state. Especially the topic of deletions, which we will discuss below in this chapter, is tricky.
- Setup **users and permissions** in the systems. In some tools, you might have to have elevated privileges for imports/exports ("System User") which not only use a license, but also might make security folks nervous, and they might add more requirements on you.

- Define the **packaging and release management** approach for your interface scripts. You should be very diligent and tidy with this so that you will be able to follow your thinking when doing changes in the future. Versioning Control Systems[54] like Git or Subversion might be a good idea to look at, even if you are the only developer in your organization for this tool — you need backups of your scripts anyway.
- Define the **requirements for the import development** with focus on operations. For example, your scripts should report back at which step of the overall process they are, and you should have an estimate of how long each step is supposed to take so that you can trigger actions if needed. This affects mostly IT ops people, admins, and end users.
- Integrate the interface into the **IT support structure** of your organization. It will become "another" thing that they have to monitor, maintain, and operate once the development is completed. You want to start the conversations with these folks early because they might have significant requirements that are easier to build in from the start (rather than doing something and being surprised that they don't want to take it over).
- Also think about how to **communicate** the changes. There should be no surprises for your end users, and you should use the communication channels that you've established in the Strategy phase. But also have a way of emergency communication in case things go South and the system might not be available or an unexpected error occurs. This could be a Teams/Slack channel or an email list of users.

Develop, test, and deploy

How to develop your interface is significantly dependent on the tool you have chosen as your architecture tool. There is no common standard established in the industry — neither for the programming language nor for the storage and export/import formats. Most tools understand widely used formats, such as CSV for import or Office formats for exports (doc, pdf, xls, and so on), but some tools export their database content in a clean XML structure. Others give you flat files or single *.bpmn files that you need to "stitch" together if you want to migrate to another tool. Unfortunately, there is no incentive for a vendor to agree on a common, shared format — quite the opposite: when using your "own" format, you lock in your customer even more. Not a great situation.

But there are a couple of things that I have seen in the past that I like to share:

- **Data transformation** — when designing your interfaces, take a close look at the definition of the data format. There are multiple number formats (double, integer) or time information might be stored as plain text. Furthermore, make sure that you understand the length of fields, especially if you have multiple sources that you want to bring into your tool for a certain object. Always choose the longest and add zeros as part of the transformation as needed (string manipulation).
- **Data cleansing**. Remove white spaces, symbols, replace empty cells with a value, etc. Look for the way how information is separated — for example, when transforming time from a string to local time, some tools replace the blank between the date and time with a "T", this might be incompatible with what you receive from a different source.[55]
- **Data Manipulation**: besides changing the individual strings and formats, you might also want to clean your data by manipulating it. This can include the combination of data (either append data of the same structure or doing a proper join of two sources — a more powerful version of the XLookup function in Excel), filtering of unnecessary data (rows and columns), as well as aggregation of data in the form of pivots or "group by" aggregations that will have multiple options for aggregation, such as mean, average, first, sum, and so on.
- **Create IDs** if needed — when you import from different sources for the same object type, define number ranges for different systems.

Depending on your tool and its development environment, you might have to use SQL and derivative languages. If you use a dedicated integration tool, it might have a visual editor. In case your tools don't include anything, I recommend KNIME, which is an open-source tool that allows visual editing of data. KNIME ships with hundreds of different nodes for anything you can imagine what you want to do with your data: from data cleansing to analysis and machine learning.[56]

Lastly, you will have to decide (also depending on your tool's capabilities) if you can import/export data via RESTful APIs, which might have limitations, or via scripts.

The example below shows how we have created a set of import scripts for ARIS in a project. You can see a main script that triggers individual sub-scripts that then created and updated objects, created models, and laid them out using the project template that was defined in the configuration phase (Technical Governance).

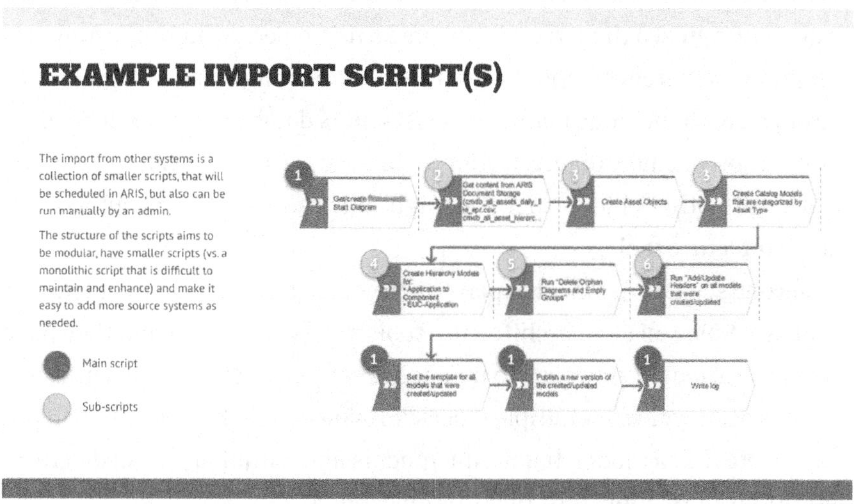

Fig. 20.4: Example import script (for ARIS)

After you have developed your interface in the Development environment and concluded the unit testing there, you will migrate the scripts to the Test environment for system and user testing, as well as load testing.

When these tests are successfully completed, you can migrate the scripts to the Production environment and communicate the new release of scripts.[57]

Sidebar: Import result comparisons and deletion of content

When developing your script, I recommend taking a closer look at deletion of objects and comparisons between imports.

The latter is pretty straightforward, you want to know what has changed with the latest import. This could include statistics ("x models/objects created, updated, deleted") or list all changes in a tabular format sorted by model and object types, as well as an area of your tool's database — see the recommended database structure in the previous chapter.

Deletion is more tricky. While it sounds simple, you actually might not want to have an automated deletion of objects and models. The deletion could be either done mistakenly — someone deleted an object in an upstream system, recognized the error, and created a new object, which unfortunately results in a different ID and the architecture tool might not recognize it as the object with the same name that exists in its database. This will result in the creation of a new object in your architecture tool ... without any of the relations to other objects that you might have added during your work in the architecture tool.

A second scenario is what happens when your script deletes content in your database. Since all good architecture tools are database-driven, they have an object-oriented approach to managing content: "create one definition, use it as occurrence in multiple places". If you delete the object, it will not only delete it in a model, but in ALL models and with that, it will delete all definitions of relationships that this object had.

You cannot fix this after the fact (in the end "deleted" means deleted) and your only hope is that you have a good backup strategy, but typically, you lose the work of a full day when restoring a backup — better than having to recreate (and negotiate) all relationships, but still not ideal.

The better way to prevent this scenario is to program the import in a way that it will populate a "Deleted" flag in objects and models. Then you can create a report in your tool that will list the deleted things in your database.

Then you can review the proposed changes and delete the flagged items manually. Or clear the flag and have a conversation with the upstream system owner why he or she should not have deleted the object or model in the first place. For example, if you would create a regulatory compliance issue that the old object had managed.

Operate your interfaces

After you have deployed your interface scripts, you need to operate them. Here are a few things to think of when doing this:

- Schedule the imports, ideally over the weekends. This might require that you have all users log out of the system, and you can be sure that some people will not do this. Therefore, you should have a policy in place that reminds them that if they don't log out, an admin will kill their sessions, so that any open model will not be excluded and can be updated.
- Do backups before doing any updates. Things can go South, and they will go South if you don't have backups. Ask me how I know ...
- As mentioned above, have notification workflows in place that will inform IT ops and admins about the status so that they can react if needed. You also want to have comparison reports and other information automatically created with each import/export.
- After the import, check the content in the import database before you merge it into the development database to catch any errors (see the sidebar about deletions as one of those unwanted results above).
  Restore the latest backup if needed and communicate this to your end users so that they can adapt their plans. You rather have users complain about not-current catalogs, than having created "holes" in your architecture that will show up weeks or months later, and you cannot easily recreate relationships or attributes. This, and the fixed weekly schedule, will buy you time to figure out if there is a data or script problem.

- Overall, the best advice is to communicate, communicate, communicate. Nobody wants to be surprised. Include not only your end users, but also be transparent to other projects and programs that you have a relationship with, as well as the various stakeholders that you have identified in the Strategy phase.

21

Technical Governance And Content Integration Summary

In this section, we took a look at the technical governance of your project. The main takeaways are:

1. Configuring your architecture tool is more than just managing users. You need to define the data that you want to capture and integrate them into a meta model before you start "wrenching" on the tool.
2. When configuring the tool, study the corporate identity guidelines of your organization and make the application "look and feel" as if it is "your application". This will help with adoption later on (and we will talk about this in the upcoming section).
3. When setting up your tool you should set up multiple environments (Dev, Test, Prod) as well as multiple databases.
4. Establish a quality/approval review process between the databases. Automate these workflows if your tool allows this, so that the participants in the workflow can get notifications or delegate tasks. Bonus points when you monitor these processes with Process Mining or dashboards so that you hold yourself honest when it comes to SLAs.
5. You are not an island. Work with other projects/programs to understand what the application landscape looks like, and what the Systems

of Truth for which data are. Develop automated interfaces to and from these systems and operate them on a schedule (e.g., weekly).

6. When developing scripts (reports, semantic checks, queries, macros, interfaces, etc.) follow good IT hygiene: always develop in Dev and promote via Test to Prod, have version control in place, and define predictable releases so that everyone knows what is coming and can adapt.

V

Process And Organizational Governance

22

Process And Organizational Governance

"What are the rules of engagement?"

Have you asked yourself that question before? Do you know that making changes in Prod is not a great idea, or how you operate that technical interface to that external System of Record? Or what you in your role as "x" are supposed to do, and when you have done it well?

What's in this section?

Well, this is the section in the book where we are covering the "process of process" — how to define the processes, roles, and organization. How to stand up those org units, and lastly how to create an enablement program that will bring everyone on the same page and create the necessary skills to succeed.

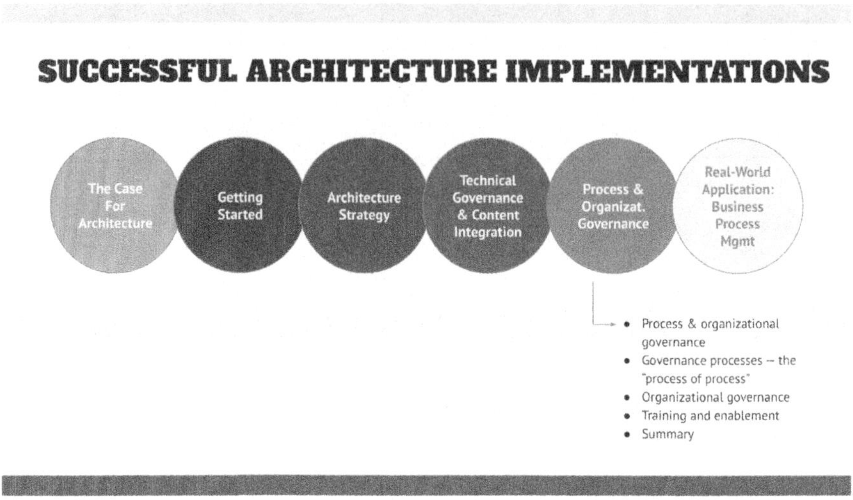

Fig. 22.1: Process & organizational governance structure

23

Governance Processes – The "Process Of Process"

Governance is a critical success factor in your architecture implementation. If you just let users create artifacts without the standards that we've discussed in the Technical Governance segment, you are missing comparisons and analysis. But you also miss out on opportunities to collaborate and create results faster in a higher quality if you do not look at the process and the people.

This chapter is looking at the first of the two, and the next chapters will cover the people side — from an organizational perspective (org units and roles) as well as an individual perspective (enablement).

Two aspects of process governance

When looking at processes for your implementation, you need to take a step back and distinguish between two different audiences and levels of detail:

- The high-level processes in which architecture can be helpful and be a competitive advantage.
- The detailed governance processes that apply to your program — the "Process of process" mentioned in the title.

Both process aspects will have an impact on the success of your program.

High-level governance processes

By now, you have a good understanding of how architecture can improve the performance of your organization and help your stakeholders accomplish their objectives (I covered this extensively in the Architecture Strategy section of this book). But **how** do you bring this into reality?

I recommend the following steps:

1. **Review your objectives and identify the use cases** that you can/will support. Some of them seem to be "obvious" (like "Process improvement" or "Application Rationalization") because they are driven by your group, who had the idea of implementing architecture as a capability in the first place. Others might be less obvious because they are driven by other people in the organization that you might not have spoken to. Think of "Risk & Compliance Management" or "Solution Development" — or the big catch-all phrase of these days: "Transformation." Also read the annual report of your organization and see if there are changes in the high-level objectives that haven't trickled down to the stakeholders that you worked with before and see if architecture can be helpful for them.
   Either way, start a list of these use cases and identify who in your organization is responsible for these topics, and add them to your list of stakeholders as described above.
2. **Rank the use cases**. There might be some "low-hanging fruit" because you have already heard that use case multiple times, or a use case is directly under your control, and you know that you have the organizational sponsorship (or even better, a mandate) for it. Others might be triggered differently: has your organization received a consent order from the regulators, or have you heard that your IT projects have a low success rate? In parallel, your CIO is complaining that he cannot catch up with new stuff (projects or technology like AI)

because his budget stretches thin "just to keep the lights on."?
Develop decision criteria for this ranking, so that you can compare apples to apples. These might be things like "degree of pressure" (internally or externally), "relationship with stakeholders", "potential benefit for the organization" (quantifiable business case), "long-standing problem", and so on. You want to rank the use cases that are solving a real business problem and have an impact, but also avoid a long "death march" that has seen other groups burn out on and that will take years to implement.

3. Develop a **pitch deck for the stakeholders** in which you outline the benefits of architecture support for their use case and how you will support them. Figure 23.1 below shows an example of a project delivery process that lays out the steps where architecture (either on an enterprise level or a solution level) can help.
4. Once you find an agreement with your stakeholder, **map out the process** in more detail (but not down to the who-does-what-when-where level of a BPMN diagram) and identify how the various artifacts of the Solution Lifecycle, that we discussed in the Solution/Process Lifecycle chapter, will fit into those. Also identify the org units that participate in the process.
This will help you identify the artifacts that an architect will have to create (and you might not have these on your radar so far and requires additional configuration of your architecture tool), but also the services that you will provide to cover the use case. Involve the stakeholder's people in developing the details of the use case and the high-level process to gain acceptance with them.
5. **Get formal agreement to the use case** so that you can add this to your services portfolio. Also **define objectives and KPIs** (and add dashboards) to hold everyone accountable and provide visibility into the contribution of architecture to the use case success.
Like I said before, articulate them in the language of the stakeholder and capture business-relevant measures. For example, to which degree have risk audits been accelerated because you now have a well-defined

process for those, and you also brought in Process Mining to show what *really* happens in the processes, and you can compare that to the documented process and show the degree of compliance automatically?

For all use cases, a well-documented and agreed upon high-level process will be a good map for everyone involved, and the dashboards will show the degree of your contribution to the success and the improvements ... this will help you in the annual budget conversations where you have to prove to "earn your keep."

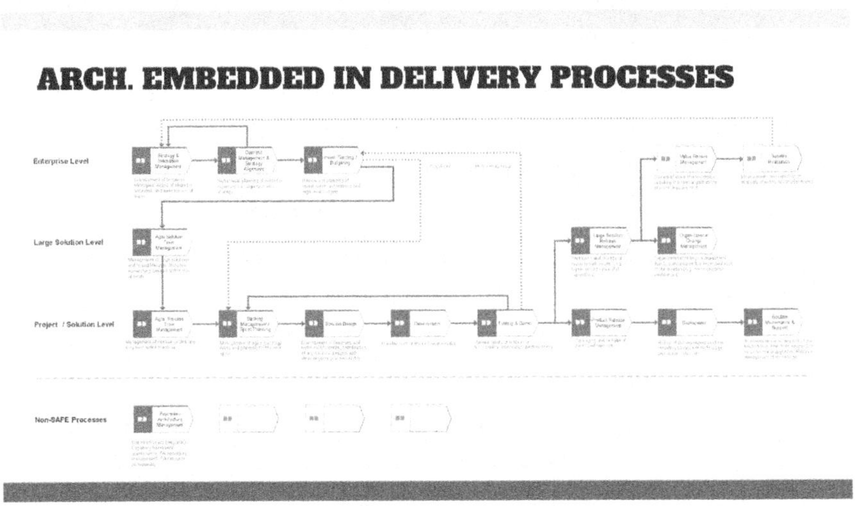

Fig. 23.1: Enterprise-wide integration of architecture in a use case

Architecture governance processes

While the high-level use case processes might be widely different from each other, the main purpose of this step is to create a network of architecture governance processes that is stable and can be extended when new use cases come up. Existing ones will then be changed.

Figure 23.2 below shows a diagram for a client that we've created in which we added the risk use case. The regulators wanted to see a regular review of the process diagrams that the business units needed to maintain to identify the risks at the lower-level steps in a process. To accommodate this use case, we defined a 30-day Review Cycle section of the high-level architecture governance process. This had the objective to ensure (and automate) the review of all process diagrams every 180 days and -as a side objective- to keep the content in our architecture tool "evergreen", so that analyses that used the process information will give the correct results.

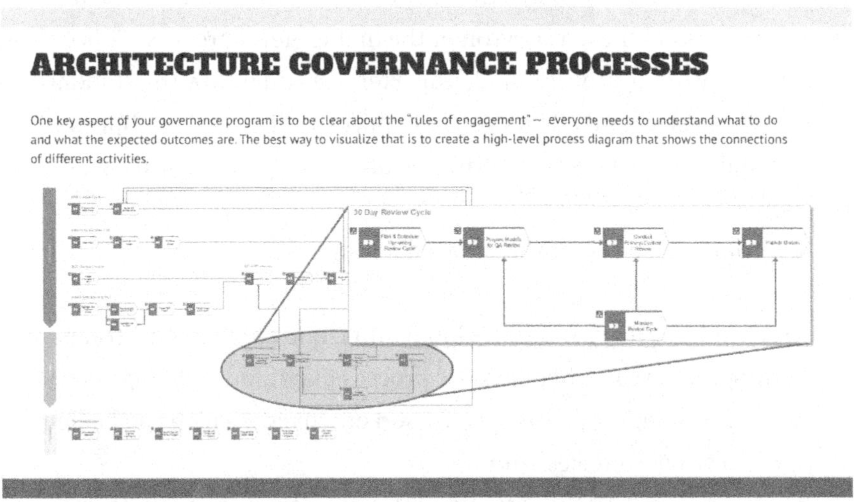

Fig. 23.2: High-level architecture governance processes

When creating this high-level value-chain process framework, you might want to think about how to structure it so that it is readable for others when it grows over time. You can see this when looking at the gray arrows in the graphic, and this structure can be aligned to phases of the Solution/Process Lifecycle or major activities in the process, like "create", "review", "automate", or "monitor".

The result will be a process network that is easy enough to understand and fits on a single sheet of paper.[58] You might want to create overlays for the individual use cases in the same way you would create End-to-End processes in a business architecture, as described in the "Real-World Example" of the next segment. Use this high-level process model as one of the key items for your communication program, and make it (and its subprocesses) available in your architecture tool for every group/stakeholder that you have identified.

Detailed governance process

In this last step of the Process Governance, you will now create a detailed (BPMN) diagram for each chevron of the higher-level process. This will be done like any other process modeling exercise and follow the standard — following the standard allows you to export the diagram as *.bpmn file if needed and use this technical format to automate processes in a workflow engine.

My recommendation for the BPMN diagrams are:

- Track at least roles, organizational units, and applications in **separate catalogs** and reuse them in your process diagrams. You might want to add more details, such as specific screens/pages, which can be helpful for reporting exercises later.
- Structure the BPMN diagrams with **swimlanes** for roles (org units) because you want to use them for communication purposes. If you have fully automated steps in your process, then change the object representation (task type) but leave the step in the existing swimlane to reduce the amount of crossing connections and make your diagram more readable.
- Apply all **Quality Assurance** tools that you have defined in the Technical Governance (semantic checks, etc.) to your governance processes as well to ensure consistence with your other process modeling activities so that your processes will fit in nicely and do not confuse other viewers when looking at them ... this cobbler's kids will have the same shoes as

the other kids :-)

- Once you made sure that things are consistent in your lower-level processes and a viewer can navigate easily horizontally and vertically, **socialize** them with your stakeholders and prepare them for the **training of your users/participants** in the governance processes. How to set up a good training will be discussed in the "Training and Enablement" chapter below.
- Establish a regular review of the processes, just like any other content in your architecture tool, so that you keep it fresh and your users will get the correct information.
- This is especially important if you create derivative content from the processes, such as checklists shown in figure 23.3 below, or if you automate your governance processes — the process diagrams should always be the master.

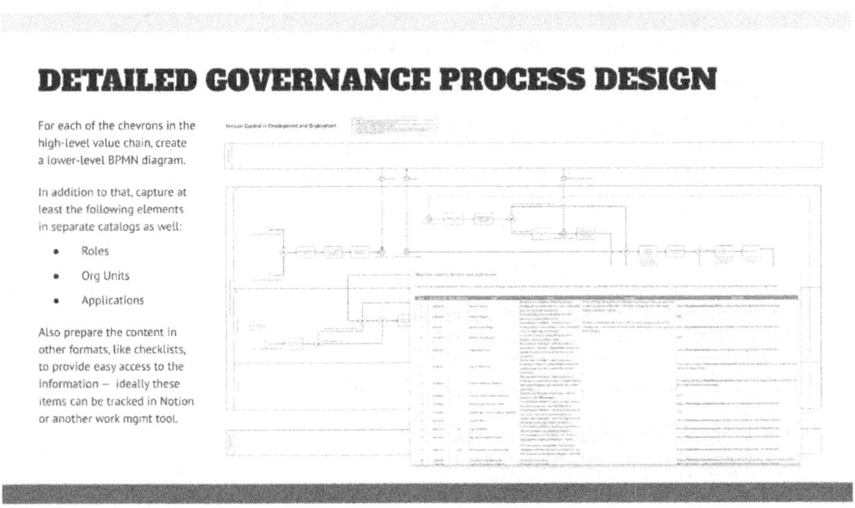

Fig. 23.3: Detailed governance process design

24

Organizational Governance Design

Now that we know what the people and stakeholders shall do, we will take a closer look at the organizational side of things. In this chapter, I will cover the following aspects:

- A more detailed look at the roles in an architecture program (which might give you some ideas on potential gaps in your processes).
- The potential implications on your technical configuration.
- The organizations that are involved in the overall governance design.
- An example of implementing an Architecture Review Board as an example of one of those organizations in a previous project.

Roles in an architecture program

Take this subchapter with a grain of salt, since your organizational setup might differ. However, what I have typically seen in organizations was similar to the roles in figure 24.1:

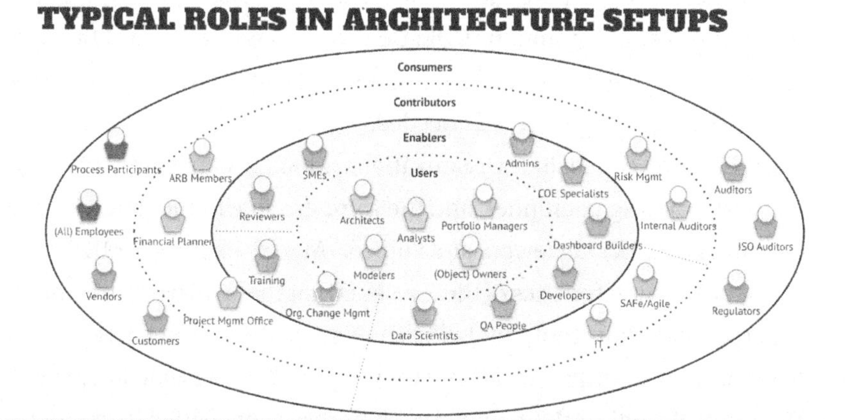

Fig. 24.1: Typical roles in an architecture setup

The high-level types of roles in programs can be clustered into two big buckets, which also might be reflected by the license structure of your tool: creators and contributors.

The creator category contains two types of roles — the users and the enablers.

- The **user** category contains all the people who do the actual architectural work and include Architects, Analysts, Modelers, but also Owners (such as application or process owners, who will make changes in your tool and approve items) and Portfolio Managers. The latter can cover different types of portfolios, such as projects or applications — their task is not necessarily to own the objects in the portfolios but to manage and coordinate the items logically, for example in the context of large transformations or for a specific business unit.
  The shared characteristic of these roles is that they all will work directly with your tool and therefore need to know the tool very well, but also need to understand their role in the governance processes.
- The **enabler** category includes different groups of roles that are "part-

time" involved in the actual work — the main group might be the Center of Excellence (COE) personnel which can include Admins, Specialists for methods and running the COE, Developers and Dashboard Builders, as well as "shared" resources like QA People and Data Scientists that will be deployed to projects as needed.
Another group can include Organizational Change Management Specialists and Training personnel, and the third group will include roles like (business or SOX) Reviewers and Subject-Matter Experts (SMEs).
The shared characteristics of these roles is that they will work in a limited capacity with your tool, but bring their specialty to the party. A Data Scientist, for example, will bring their data modeling tool and provide Python/SQL coding skills that are not necessarily needed to create the artifact for the use case.
However, all the enablers require at least the "doer" knowledge of what you defined them to do in your governance processes, but not necessarily need to understand the "rules of engagement" or how things like delegation will work — they are "just" playing their role.

The contributor category includes two sub-categories: (hands-on) contributors and consumers.

- **Contributors** typically include "content roles", such as decision makers or specialists in the form of Committee Members, Budget Planners and Controllers, or the Project Management Office.
  It also includes roles that help running the system, such as IT people who operate the IT technology for you (backups, upgrades, interfaces, etc.) or (Scaled) Agile people who will help to manage architecture projects. Lastly, the Risk Management people and Internal Auditors will be players in your program.
  The main characteristics of these roles is that they mostly will only read the content of your architecture work, but they will have an input on how to make that content better, while not directly working in your tool — they will most likely work directly with your users and the

Reviewers/SMEs where applicable.

- The last group are the **consumers**, which are in theory everyone who has the need to view the architecture content. This can be the Process Participants who perform in your processes or — even wider defined — everyone in your organization.
  But it can also contain external roles (red in the diagram above), such as Auditors (risk, ISO, etc.), Regulators, Vendors, and also Customers in case you provide technical services and integrations with them.
  The main characteristics of them is that they just need to be able to understand / view the content that you create. One thing you want to consider is the format of that content — not everyone is familiar with reading diagrams or navigating through a sequence of forms to get to the content of interest (or they might wear gloves on a worksite or do not have computers/tablets available), so you might want to think what alternative ways of giving them the content might be ... the good old printout might just be what is needed. Alternatively, think about how that content could be integrated into their workspace (machinery, workflows) or embedded in their environment (augmented or in real life, such as signs).
  Unfortunately, most of the current tools fall short outside the computer setup, but I have high hopes that the vendors will wake up and see the potential of other representations.

Potential implications on your technical setup

As you have seen, the different roles have widely different requirements on your tool configuration, and there are a few aspects you need to consider next.

The first one is to translate the requirements into the various areas that your tool allows you to configure. These are typically structured in these categories:

- **Licenses** — and to be more specific, user licenses, since there might be other parts of your licenses that you will need for parts of your tool stack that are not relevant for users. For example, when you look at typical process mining license setups you will see user licenses, but also connection and data processing licensing and other fees that make the license setup as clear as mud (and I guess that is intentional, just like with your cell phone contracts :-(

Licenses, however, typically constrain areas of functionality of your tool. Can your user view or edit items? Does she have admin access, or can she use specialty functionality in your tool, such as simulation or process mining? The structure of licenses is different from vendor to vendor and might overlap with the next category of privileges.

- **Privileges** describe what a user is allowed to do. While a certain license might be required to access a certain part of the tool, some tools also have the concept of privileges, which will give you the option to further define what roles can and cannot do.

  An example would be an Admin — they need an admin license to even see the admin section of the tool, but if you want to be more granular you can further define different types of admins with privileges — can the role manage users, do backups, etc.? Or do you specify User Admins for each main org unit in your program so that they can administer their users, and you do not have to do that centrally?

  Or you have Developers who typically also use an admin-type license, but you want these roles to be only allowed to create and manage dashboards and reports, but not the overall installation and its configuration.
- Then you have **permissions** — these are used to define access rights to specific areas of your architecture. The general rule that I apply in my projects is that everyone can view (and reuse) things, but can only define items in areas that they are assigned to.

  This means that I set up structures in the tool that allow an org unit (user group) only to edit "their" content (an HR group can only edit the HR processes or IT can only edit applications). Everyone can see what these groups create so that, for example, everyone only picks from a list of IT-managed applications and doesn't "invent" their own.

However, don't be surprised how "creative" humans will be if they don't find what they are looking for — I have seen tons of created objects outside the official areas, for example applications, risks, or controls. That might be because there is inconsistent naming of things in your organization and the users just didn't find them, or they were not aware that they are not supposed to create those items — this can be a failure on the governance process side or a lack of training that you might want to improve going forward. In any case, you need a means of finding these scenarios (reports) and consolidating them, but also a proper permissions setup that will allow you to do this (e.g., all "official" apps will live in one specific area that only IT can edit, and your report will find all applications used in the database whose object definition exist outside the sanctioned area. You also might want to think of automating these reports and trigger workflows to mitigate the inconsistencies.

- Lastly, you might have specific **tool configuration items** that are applied to roles. For example, your risk and compliance users might want to see a specific indicator if a subprocess in a process diagram includes risk-relevant information without the need to open another subprocess diagram. This information might not be relevant for other users and clutter the interface so that you might want to configure that in a different configuration file(s) — in the ARIS example from the Technical Governance segment this would be done via filters, templates, and the portal configuration.

My recommendation is to create a matrix as shown in figure 24.2 below. When you do this, keep in mind that it includes licenses/privileges/permissi on structures for all tools and that you standardize your roles into groups that are consistently applied over the whole user population.

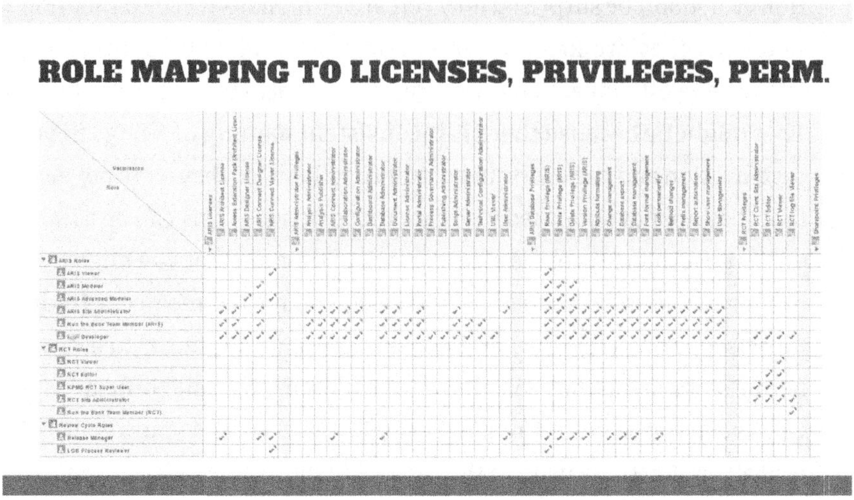

Fig. 24.2: Role to configuration definition mapping

This can be done by creating user groups that are aligned to the role groups. You might have a user group of "HR modelers" and "IT modelers" and both user groups refer to the definition of the "Modeler" role group and the users in both user groups can do the same thing and the only differences are in the permissions.

And just to have it said, try to avoid granting licenses, privileges, and permissions on an individual user level. This will rapidly become a user management nightmare to maintain, and you might fall out of favor when your security people will do an audit of your setup and see that you have a million different configurations. The worst-case scenario is that you cannot quickly and logically explain how you consistently applied the three items to your users.

Governance organizational design

Now that we have defined the roles and have them assigned to the tool features as well as to the governance processes, we need to think about how we structure the org design for your program.

Do you remember the "Content — Governance — Adoption" graphic from the "Capabilities" chapter above? Those three areas make or break the success of your implementation, and you need to establish governance functions so that the roles in your program (mentioned as Architecture Community in figure 24.3 below) can do what they are supposed to do.

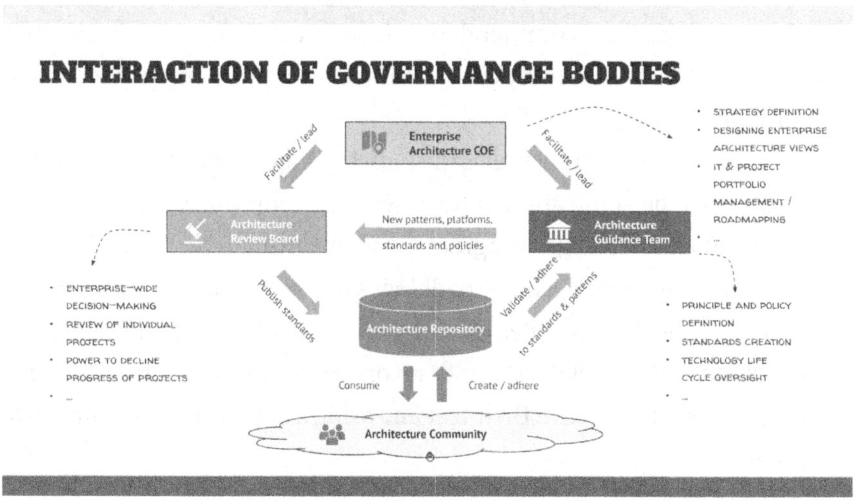

Fig. 24.3: Interaction of governance bodies

Being fully aware that I might sound like an old Middle School teacher, but we will adapt an old concept to the organizational governance of the program.

- The main idea of an architecture program is to create content, and you will need to give direction and support — your **Executive Branch** function.
  This is mostly done by the Enabler role group above and might be structured as a Center of Excellence or as a "Guild" if you follow the Spotify model of organization.[59] It doesn't matter how you call it, as long as this group consists of people "who really care."

- The second function is to define and maintain the shared rules (the **Legislative Branch**). This part-time group should be filled with architecture people who are a bit "nerdy" and will be able to analyze standards, reference content, framework, etc. and apply this to your program.
  This reflects the fifth major architecture capability "EA Methodology" and will have a direct impact on the configuration of your tool. But, as always in life, my recommendation is to do this in moderation so that you don't kill the creativity of the Users or overload them with too much formality.
- The last governance function is the **Judicial Branch**, the "referees" when it comes to the application of the rules. Typically, you see Architecture Review Boards in larger programs, and unfortunately, I have seen too many of those who became roadblocks for innovation by insisting on following the defined rules and standards "to a T". They have become the "Committee of No" and since no one wants to play with unpleasant people, people vote with their feet and bring a big risk to an architecture program.

As figure 24.3 above shows, the three functions work together, and you will have to find the right balance between them and the actual community that they are serving (just with the three areas of successful architecture implementations). Unfortunately, there is no "one size fits all" approach, and you should define the players in these groups, as well as the names of the groups, according to your organization's culture. Do not be "too esoteric" because that will reduce the adaptability of your program — nobody wants to be associated with the perennial loser in your favorite sport.

However, you can make the best out of this situation and be as transparent as possible in the design of these org units (map out their processes and define their overall objectives), and the communication about and from them. And just like any other program team, there should be KPIs that measure the performance of the three branches shown in dashboards, and if something doesn't work out, then change things.

Implementation of a governance org unit

The last step in the org governance section now is to set up the org units that you have defined. This can be done in two ways:

- The onboarding of regular user groups with a standardized process and training, or
- The specific implementation of a "general"/"shared" governance or unit, such as a COE or an ARB.

I will cover the first scenario in the next chapter, when we discuss what to do to create a training/enablement program. I recommend adding the topic of standardized onboarding of new users or user groups to the program to the governance processes of the previous chapter. This allows you to rationalize the roles that you might need, specify what you expect them to do, and how you can support these activities with systems, automation, and visibility (dashboards and reports). Remember that they are most likely in the contributor role category from above and do not understand completely how the show is run, but rather just perform their part of the bigger picture.

For the second scenario, I cover an example of our favorite governance entity — an Architecture Review Board.

A few years ago, I worked on a large transformation project with a global organization and helped them define an architecture capability and implement it following this approach(-ish). However, when it came to defining the organizational governance they were not happy about having an ARB, since they already tried this a few years back, and it was the "Committee of No" and everyone involved hated it.

I was able to convince them to establish the three types of governance orgs that we just discussed, and they decided to give an ARB another shot, but wanted to have a low-touch approach to it. Based on this, we developed a four-step approach to the maturity of this org unit and also established clear rules on how to run the ARB.

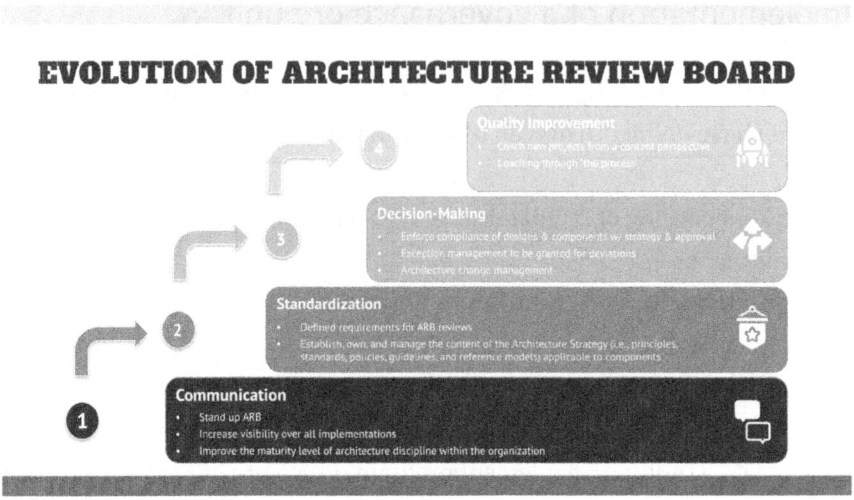

Fig. 24.4: Evolution of an Architecture Review Board

The four stages of the ARB evolution were:

1. **Communication** — providing a forum where the different parts of the organization that were tasked with implementing the transformation could meet regularly and exchange information about what the individual projects do, what their objectives were, and what they were implementing from a technical perspective.
   We defined a core group of ARB members and developed a schedule for individual projects to present. Everyone in the transformation program was allowed to attend, and we also streamed the meetings via Teams, so that remote workers could collaborate. This allowed, after a period of curiosity where people checked if the new ARB was not just like the old but rather a true new approach. This enabled to exchange information and, over time, lively discussions about the right way forward and what to use from a tech perspective.
2. **Standardization** — the first step of my engagement with that client was to do an EA Maturity Assessment, and it became clear fairly quickly that they did not have any standards in place and groups just saved

whatever they needed wherever they saw fit. They also used different notations (or none) and did not have a shared understanding of which artifacts should be included in a Solution Design.
As part of this phase, we defined standards (just a few) and prepared for the purchase of an architecture tool, which would then come with a whole set of technical governance and would further standardize artifacts in one repository for enterprise-wide analysis.

3. **Decision-Making**. The phase that everyone was fearing to bring out the worst out of Board members, and everyone was afraid that it would evolve into the "Committee of No" again.
But you know what? We never reached this stage. One premise of the ARB set up was to stop at a level where everyone would be comfortable to move on. And the client was happy with the collaboration that they had so far, and decided to deal with exceptions, changes, and roadmaps in the collaborative format that they had established. Decisions were made by all involved (even the "non-core" members in attendance) and that worked out very well.
4. **Quality Improvement** — In this stage, the ARB characteristic changes from the place where you have to do your "sing and dance" to a true helping organization. The ARB helps new projects to get off to a good start by sharing content knowledge and help them through "the process". It transforms into the "Committee of Know".

Evolving the participants and the committee was a very delicate dance that required a lot of communication and clear guidelines to build the trust needed to proceed. And at any point the committee was allowed to disband itself if the outcomes would not meet the expectations.

For the first phase, we established these "rules":

- Architecture review board will have regularly scheduled bi-weekly meetings.
- Project teams will request to be included in the agenda in advance.
- Project teams will provide the appropriate materials for each checkpoint.

- The architecture review board will determine if a project is exempt from the review process.
- Architecture review board may require project teams to follow up on questions or comments after meetings.
- No approvals/rejections will be incorporated in stage 1.
- The scope of the architecture review will evolve to meet the IT and business needs.
- The Architecture review board will not be responsible for managing the project portfolio.

As you can see, we defined steps in the development process when projects should check in with the ARB (visualized in figure 24.5 below).

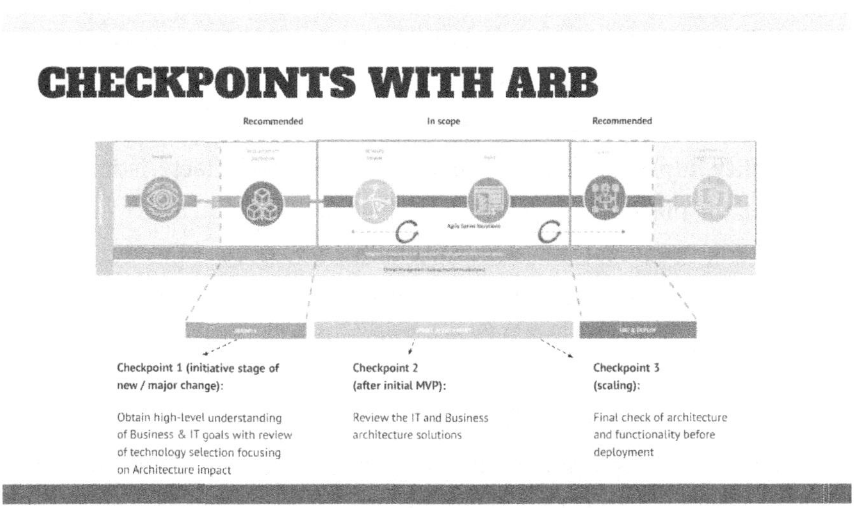

Fig. 24.5: Checkpoints with the ARB (project example)

Having those clear defined rules for the ARB operations and following up with them transparently allowed the ARB to evolve to a level that "felt right" for the organization. This created the trust needed to bring all projects into the fold without having to set up organizational "orders" that let the first attempt fail.

When implementing governance organizations in general, you want to ensure that you understand where the people are, what their motives and expectations of joining such a group are. Then tailor an approach similar to the ARB example above that allows building trust while keeping as far out of the project's businesses as possible, and have them have the responsibility (and authority) for their implementations.

25

Training And Enablement

The last element of the Organizational Change Management activities (besides communication and org design that we covered before) is the training and enablement area. You want your users to be up-to-date with what they need to learn for their role in the architecture program.

The good news is that most organizations already have a training program in place, and I highly recommend to simply follow their guidance and processes. Setting up a global learning program in multiple languages and for different products/learning areas is no small effort (guess, how I know this ;-).[60]

The following content describes what you should do in case you do not have a learning group. You can also use it to give additional input and ideas in case you already have something established, and you think it could need some fresh ideas.

What are the steps for setting up an enablement program?

Setting up a program from scratch is typically structured in three phases:

- Design of the program
- Development and training delivery
- Management of the training program

In addition to this, I highly recommend looking for a Learning Management System (LMS) that you will use for managing online and in-person content. These tools allow you to not only manage the content of your classes, but also allow sending out surveys after delivery, do reporting, etc. — a lifesaver in some situations. However, the implementation of an LMS is out of scope for this book.[61]

Figure 25.1 below shows the three phases in an overview:

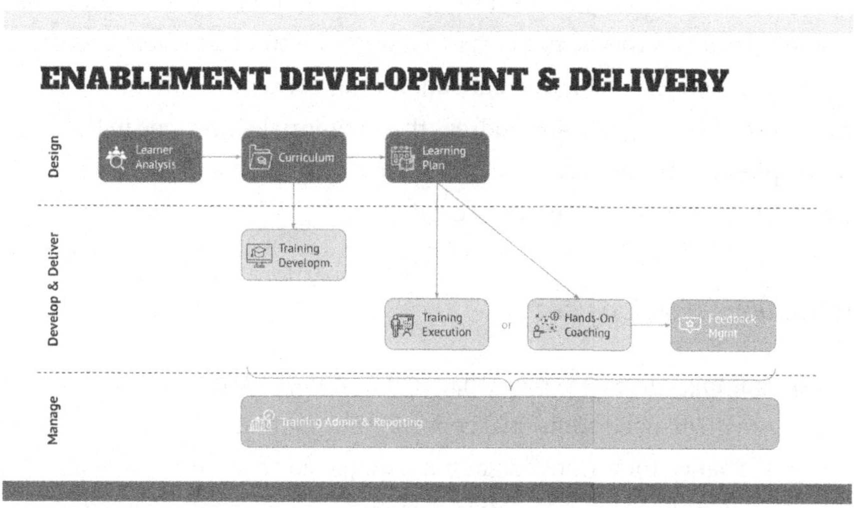

Fig. 25.1: Three phases of enablement development and delivery

Design phase

The first phase is the Design phase of the program. In this phase, you have to look at the learners that you will cater with your program. These are the roles you have identified in your process governance, the groups that you identified in the organizational governance phase, and the various stakeholders that you determined in the strategy phase.

Learner Analysis

Once you have identified the roles, you start with the learner analysis. For each role, determine what they need to learn (the easiest way is to run a report on the governance processes that you have already modeled — this will give you the steps that users will do in which process and which tools they will use for that), and what their previous knowledge of the subjects are. For example, have your users already worked with similar tools to your architecture tool, or are they complete newbies?

In addition to that, you need to determine their change readiness and where they are on the change curve that we've discussed earlier. This could mean that you will have to create additional content for them to "bring them into" the program and address their potential objections in the form of supplementary material and micro-learning ("why do we do this?"), which are typically not found in a curriculum.

Curriculum development

A good training consists of more than tool training ("Which button do I have to press?"). You need additional conceptional content ("What is BPMN?") and governance material ("What am I supposed to do in my role as x?"). You also might have to add additional communication content, as discussed above, to support the adoption of the program.

This will give you the raw list of things that you need to teach, and you should also include resources that you find elsewhere, such as training offerings from your tool vendor, or organizations that you are part of, such as TOGAF. There is no need to reinvent the wheel, and you take away the effort of having to maintain that content every time when the software updates or a new concept is added to a framework that you follow.

The next step is to map training content to the different roles that we've seen in the previous chapter. Typically, you see different levels of expertise needed that can be mapped to levels like "Foundation", "Basic", and "Expert":

- **Users** need the main content in detail (tool usage, governance, concepts)
- **Enablers** might need different content depending on their role as admins, developers, or data scientists. These roles are also "heavy users" and might need the same level of deep detail, just in different areas (such as the administration of the tool, or the APIs).
  However, there are also less involved roles in this group, such as reviewers and SMEs — these might need a "less heavy" content in the form of overviews or stripped-down content that just covers what they need to know. An example of this could be automated governance processes — they don't need to know how those will be set up in the tool, but only need to know what they have to do as a participant in these processes.
  Either way, all of these rules require governance and concept content as well.
- **Contributors** are less involved, and some roles in this category have viewer status or do limited data input. In any case, their needs are more around overviews and their role in the governance processes. From a conceptual perspective, they might need maybe a bit of knowledge (a reviewer or SME should be able to read a BPMN diagram for example) or might not need anything (like IT people who operate an on-premise installation).
- The last group of **consumers**, which could be internal (e.g., process participants) or external (e.g., auditors) typically require just high-level, overview type content, such as "where do I find content that I am

interested in following the structure of the tool". Typically, you don't have any special content for these roles besides overviews and basic "where do I have to click" information, and I have seen organizations that build dynamic in-tool walkthroughs for this group of users.[62]

This shall provide you enough information to lay out the **content in a structured way**. One visualization that I like is mapping it out as a subway map, where every logical piece of content is represented as a stop. You can also see where related content is connected.

In the example below, you can see that process modeling (activity in the tool) not only requires knowledge about the notation (concept/BPMN), but is also connected to other activities like simulation, exchanging diagrams with process mining or the automation component.

You can now define also which subway lines are for less involved roles (the overview line) or what the foundational content is, and you can point your users to just the content that they need.

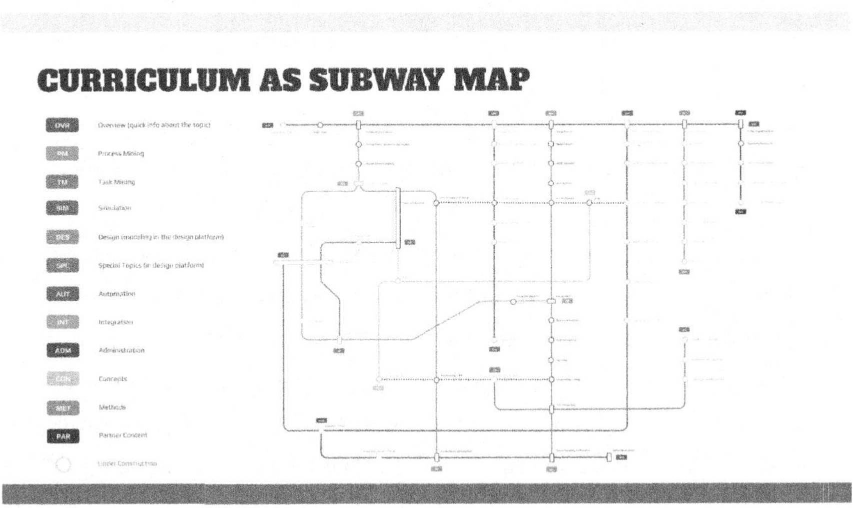

Fig. 25.2: Curriculum visualization as a subway map

One thing you definitely should do is to define **learning plans** per role — the sequence of subway stops and where a learner might have to switch lines. The successful completion of a learning path can also be the prerequisite for a **certification**. If you use an LMS, then you can use their functionality to structure things and also upload required material that someone needs to have read before they start a course.

Then you define the **format of the courses**: will you provide instructor-based classes, self-paced e-learning courses, micro-learning, brown bag sessions/events, or any other form? This will determine what effort and which tools you might need besides a presentation program. When designing your courses, keep in mind the different learning styles that different people have — the most common ones are visual, auditory, read/write, and kinesthetic (doing/moving/touching).

Finally, you have to consider is if you require learning labs for your software stack (in almost all cases, the answer will be "yes"). Then you need to determine how those can be provisioned — does the tool vendor have those? Do you want self-service, or do you need to provision them manually? How long do they need to be made available (the cost can creep up pretty quickly)? And so on. My recommendation is to automate as much as possible and hopefully, you have chosen a tool that allows multi-tenancy so that you have only one installation, but you can provide multiple instances for the learners on that one installation. Or can you provide local installations with a temporary key?

BUT: for all of these thoughts on creating a curriculum (and also the learning plan and delivery below), keep in mind that you need flexibility. Be prepared to change if you realize that things don't work out and your learners don't reach their goals.

Learning plan

The learning plan has two aspects — an external facing part that has the learners in focus and the internal facing part that looks at the update and ongoing maintenance of your content.

For the external facing part, you need to look at your roll-out plan and when you want to provide which capability/service to whom. This will result in "learning packages" that you will provide over and over whenever a new group is onboarded to the program once you got things setup. For the initial roll-out, I recommend having a "soft launch" of some well-meaning volunteers that understand that things might not go well in the beginning but also can give you relevant feedback.

Once you have launched your program, you should think about how you will keep the content current. Initially, this is not a problem because it is a very fast-moving area with an "all hands on deck" attitude, but you would rather not keep that pressure on forever and rather install a process that establishes predictability when new or updated content can be expected. As a rule of thumb, I would review the content and potential feedback at least twice a year, depending on the availability of resources that can work on the courses.

Whatever you decide, communicate clearly and openly what the plan is — a wiki that holds the training offering descriptions and a calendar that shows the internal and external perspective should do the trick for you in the beginning. Once you mature, this can be done by the LMS or your work management tool.

Develop and deliver phase

This phase is the "bread-and-butter" phase of your program, which hopefully goes into a steady state quickly.

Training Development

Depending on the choice of your course format, you will have to do different things. You need to collect and sort the content of a course and determine what the best way of teaching that content is. Is it a lecture with exercises? Do you need interactive features like quizzes in your e-learning courses? What is the mix of learning offers — reading, watching a video, interacting with the tool or course, exercises and knowledge checks?

Whatever you decide, the key is that your team has a shared process of how things are developed, and you have defined how you do version management of the various assets that you will create — screenshots, videos, etc. You also need a process for keeping these current or when you retire an asset and where you have used it in multiple courses that then need to be updated. Initially, a shared drive might get you off the ground, but when your enablement program grows, you might want to take a look at an asset management tool to help you with this work. Some of these tools auto-update multiple presentations when a shared master slide is updated. Ideally, you have a workflow management tool for this, that keeps you informed about the status of the training development so that you can determine the release schedule of a course and plan the effort of creating the material.

Once you have released the material, you need to have version control established so that you can roll back a class if an update is not correct, and you can answer the question of "what has changed" in the courses. This allows a learner to determine if they have another look at the updated version of the course.

Training execution

This is the moment when the rubber meets the road. You release your training and people enroll — either in in-person classes or in e-learning. In any case, you should keep track of your learners and have them all registered in one place (you guessed it, in an LMS).

What you need to provide your trainers with is the list of all the material that they need for executing the training and the meta information, such as passwords, keys to training rooms, catering, etc. You should have backend people stand by to be able to either support the trainer (her main job is to train the learners and not organize things) or to troubleshoot when issues arise. Provide a checklist for everyone involved in preparation of each training session as well.

In the case of e-learning, you want to do the same, even though the support might be more of technical nature, or you struggle with the LMS.

In either case, you would like to take a closer look at your LMS maintenance and upgrade schedule, so that your classes or releases of new/updated content won't be interrupted by these housekeeping items (which are necessary).

Hands-on coaching

While the regular training delivery is mostly standardized, hands-on coaching is a different form of learning that is highly customized to the learner himself. It is by nature a very communicative form of enablement that requires an individual training needs assessment and execution, which might include on-the-job instructions but also might reuse existing training material from different courses as reference, instruction material, or prerequisite for a mentoring session.

Regular training has a defined catalog of deliverables and when you have checked-off all items of the list, you will have delivered a training at an acceptable quality at minimum. This is not the case with hands-on coaching, and not every trainer that you might have available can do this. I would seek people who have good communication skills, profound experience and a high emotional quotient (EQ). You also might have to accommodate the process that you want to train, which might not always be possible.

If you cannot accomplish these specialties, I would be hesitant to offer coaching as an enablement method, and rather provide enablement through the standard material ("hand holding").

Feedback management

After each delivery, you should collect feedback from the learners so that you can improve your content, but also rate the delivery by your trainers in the case of in-person learning. You will also learn about the tools and might collect feedback for the vendor as well (or requirements for the next phase of your implementation program).

Ideally, this is done in an automated fashion — your LMS will send you surveys, or a survey is included in the course itself, and learners will fill out the forms. The tools now should calculate the results and provide them in an easy-to-understand structure, such as the rating of a trainer by class over time.

However, the result might be a bit messier: you could receive no feedback (someone just clicked something in the form to "get over with it"), or you collect feedback on paper, or you get unstructured emails. Whatever you get, you should be grateful for the feedback and follow-up with the learner where needed (besides the automated "thank you" message).

The core aspect in this phase is to manage the process (you bought an architecture tool that allows you to model processes for communication!) and have a structured intake and analysis of the feedback that you receive. You want to make sure that you get the feedback in a timely manner, and you can add additional context information to it so that you can explain outliers months from now if required.

Manage phase

Once you make it to this phase, you have solved the big problems and your operations go into steady state. You receive feedback and the volume of things to be changed or incorporated shall be significantly reduced compared to the initial days of your enablement program. The setup of courses and learning plans is done, and the only dynamic thing that you need to stay on top of are the in-person classes and events — but even for those, you can use your LMS and have your learners register for the class so that you have the admin aspects covered (enrollment, attendance, completion, certification, etc.).

The focus in this phase is the reporting of activity and the accomplishments of your enablement program. I recommend following the steps that we discussed in the "Create visibility into your architecture program" which obviously will cover learning-specific KPIs. What you want to measure is not only the usage (how many enrollments, how many active courses, etc.), but also the "usage" of the courses — how many courses and learning plans don't get finished and abandoned?

This can give you the "unspoken feedback" on the quality of your training offering, and you should reach out to the users and chat with them (and encourage them to continue). Hopefully, you get the right feedback to improve your courses or remove roadblocks for your learners.

Additional support measures

Enablement is not done when learners have gone through a class successfully, but you need to provide ongoing information to them when you update your solutions to a new version (aka capability configuration) or you add new people to the program — nothing creates a worse impression than an unstructured learning offering and people cannot find their way around.

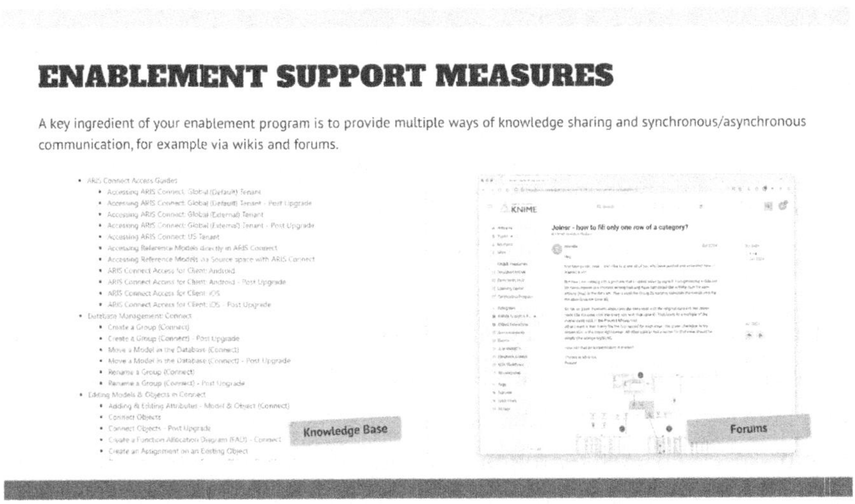

Fig. 25.3: Enablement support measures[63]

You can help them with the following (in no specific order):

- Create a **knowledge base** that can be used as a reference. This is best done in a wiki that allows versioning and version comparison (what has changed) on every page.
- Update your courses on a **regular schedule** and make it clear which one has changes, what the changes were, and what the release date of the course is. Typically, the Learning Management Systems have a card representing the courses in overviews; add the date and maybe an indicator for "new" on those as well.
- When designing those **overview pages**, make it clear what is content for an older version (if you need to keep it), and also allow for filtering by target learning level, role, etc.
- **Communicate regularly** — announce updates via a mailing list to all active users (that is a standard functionality in LMS) so that they know that they are currently enrolled in an "old" course.
  Furthermore, create "release notes" for your new/updated courses. In one of my previous employers, we used an AI-driven tool to create short videos every month that we posted in a "What's New?" section of the LMS, but also reused them on LinkedIn and other places for marketing purposes. Remember: you are not only "selling" the solutions that you implement, but also the progress and value of your architecture program!
- Provide a **real-time communication** platform. This can be a Teams/Slack channel, or a forum. In either case, make it course-specific and also moderate these places. You don't want it spammed, and you also might want to automatically close conversations after a certain time or if they become "stale". However, these posts should still be available for users view-only.
- Have **community rules** in place, which lay out the expected behavior (and what happens when you don't conform), but also provide links to the "read me first" resources that you have created, such as overview courses, links to your knowledge base, etc.
- **Establish a way to provide feedback** on courses. What I've done in the past was using tools like Asana[64] to allow users to create tickets if they

found something that needed a change, which then created a ticket in a board that the learning team then triaged by priority on four levels. If it was a level 1 or 2 critical error, we updated the courses immediately, if it was a level 3 or 4 error, we added it to the long list of changes for the regular update (twice a year).

In addition to this, every course also had a feedback survey that was sent out automatically by the LMS when a learner completed a course or an instructor delivered an in-person class.

But also keep in mind that receiving no feedback is feedback. There might be a lack of engagement with the training, or your delivery does not meet the learner's expectations. This is a situation that should make you think.

26

Process & Organizational Governance Summary

I think it became clear that you need to do more than just handing out licenses to a tool when defining the governance for your architecture program.

1. The first step is to define the processes for your program, which are aimed into two directions: the higher-level embedding of architecture activities in use cases for your organization. This will create acceptance and show the value contribution of architecture to the overall success of the organization.
   Use this to build alliances and to improve the maturity of your organization. Once you are past a certain maturity level, people will appreciate the excellence approach and wonder how things could have been done differently in the past.
2. The second step is to integrate these use cases in an architecture playbook, which is composed of architecture processes and additional information. Here you will define the detailed steps, roles, artifacts, and systems that are needed for your content. This allows you to define the organizational configuration of your tool later, and it may act as input for your training design as well.

Be aware that these two steps are not one-and-done activities, but you want to review and adjust them on a regular schedule — just like any other content in your architecture.

3. Once you have the "rules of engagement" (or "process of process") defined, you need to define the organizational governance. Start with identifying the roles in your processes and see how you can group them.
4. A good structure for your governance entities is the concept of the three branches of government applied to the program. The key success factors of an implementation (Content — Government — Adoption) are matched to groups "who care" in the function of the executive, legislative, and judicial branches.
5. When implementing these org units, be very culture-sensitive to your organization. Define them as "humble servants" and not as a "career path", and hold them accountable to what they do and communicate clearly.

   A stepped implementation approach might be the right approach for you as well.
6. Enabling all people in your program is crucial. Keep in mind that the different roles have different training needs. This covers the depth of knowledge required (some roles just might need an overview or a brief introduction, while others need to be on an expert level), as well as the content of the training.
7. Your enablement program should follow the three phases of Design, Development and training delivery, Management of the training program. Support the development, delivery, and management with a Learning Management System if possible, so that you can collect feedback, update your content easily on a predictable schedule, and enable reporting on courses, participants, events, etc. more easily.
8. Good training will include at least three different areas (to different degrees): conceptual knowledge, governance knowledge, and tool knowledge. A visualization of your curriculum in an easy-to-understand way, like a subway map, will be helpful for your learners. The format of the training should accommodate the learning style of

the learners and be aligned to the various role groups that you have defined (users, enablers, contributors, viewers).

9. Implement additional support measures that will allow roles to get knowledge they need outside of formal training (via forums, knowledge bases, etc.). This shall be supported by a clear communication from the training team.

VI

Real-World Application: Business Process Management

27

Real-World Application Of Architecture

Wow. You made it through the theoretical parts of the book. Congratulations.

And now you wonder, "how do I apply this in the real world now?" It seems to be a lot of work and daunting, so you might be discouraged and forget the whole idea. Please don't. In this section, we will take a look at the real-world application of one slice of an architecture, and you will see that you can start small with one aspect and take that as the foundation to grow your architecture — not only in content, but also across your organization to establish a shared "view of the world." This allows you to not only find improvement opportunities, but also to implement them in your process execution.

What's in this section?

We are looking at one use case of architecture (Business Process Management) and I walk you through the main steps that you will do to get to your desired results. In this segment, we will:

- See how to structure your Business Architecture in your tool.
- How to create a network of systems that provide content for your architecture efforts.
- How to import existing content into your architecture tool

- Investigate how you can analyze your processes using a "base analysis" and process simulation after you have created your architecture descriptions/models.
- Take a look at other forms of analysis in the larger context of process intelligence.

Ready? Let's get started.

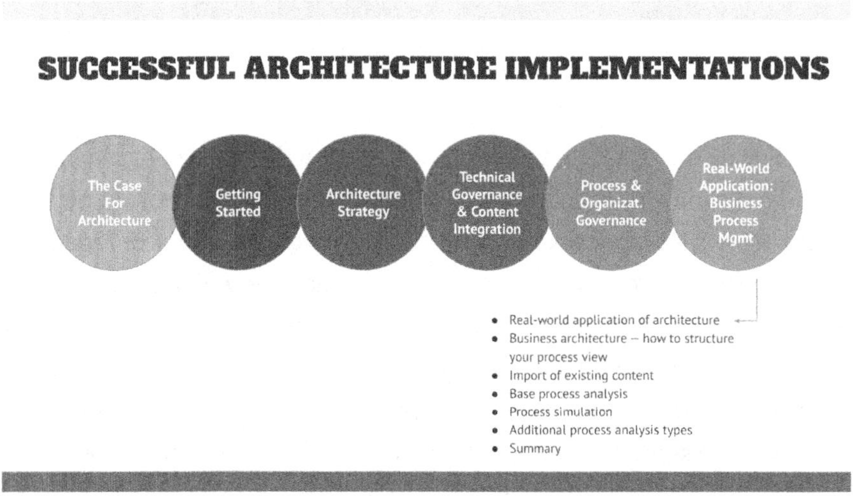

Fig. 27.1: Real-world application: BPM structure

28

Business Architecture - How To Structure Your Process View

Setting up a structure for your architecture can be challenging when you want to include the demands of all stakeholders and cover all use cases. For some architectural views, it is easy – applications typically come in "flat" and might just have one level below in your meta model (for components and other viewpoints), or risks and controls are typically already categorized before they are imported into an architecture tool.

Finding the right structure for your business processes can be tricky, though – various demands and opinions might come together that conflict with each other. On the one hand, you have the business units who want to improve a process 'soup to nuts', while others have a specific focus on regulatory compliance, and others are even looking at processes from a functional perspective. Or, in some projects, all three aspects come to play, which obviously can lead to confusion or short-term decisions that will haunt you long after the project has ended.

Which views should I look at?

Let's take a step back and take a look at where processes are in the architecture picture. Processes are part of the Business Architecture view in a larger enterprise and are connected to other views, such as applications, data, and technology (see, for example, TOGAF's ADM that shows these views in an overview). Other parts of the Business Architecture are organization and risk & controls – typically represented by people that business folks in the lines of business work with.

Business processes are the most tangible of an architecture that a regular business person will encounter – in the end, everyone in the organization performs steps that were (hopefully) designed to create an output that can be sold to a customer (a product or service). However, when it comes to describing processes, there is not a consensus on how this should be done. Some people want to have a process breakdown, while others look at lower-level steps and how to improve them (typically these are the Six Sigma[65] and Lean[66] folks), while others look at a product-oriented sequence of steps, and the last group is interested in larger End-to-End processes and how they can be measured and improved.

This chapter will bring clarity to this dilemma and will help you to set up a structure in your architecture tool that allows you to support all of these use cases, while still providing a consistency that is needed to do a structured analysis or measure the performance of the processes in scope. At first, we will look at the functional process view in this chapter.

Functional business process view

The functional process view contains the logical decomposition of business processes. It is aimed to covering a single function of an enterprise, such as "Finance". Typically, you see these broken down into 4 or 5 levels (see for example APQC's Process Classification Framework), and you can identify the main stakeholder or owner of this area – the CFO for the Finance area, for example.

When you think about the functional view, think about a big map that shows all the information, but allows you to drill in so that you can see the details better:

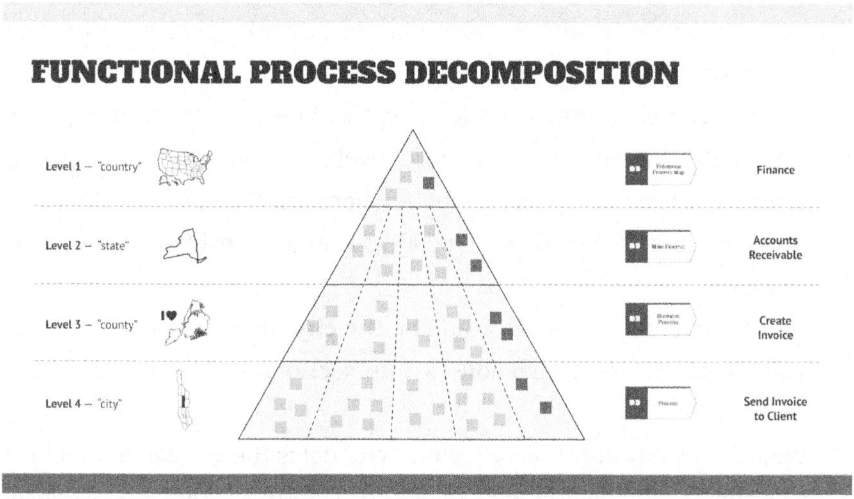

Fig. 28.1: Functional process decomposition

- On the first level, you see the **country** in its entirety (say, the US). It is one Value-Added Chain model that includes objects like "Finance", "Marketing", "Human Resources", "Procurement", "Logistics", etc. (*Enterprise Process Map*).
- When you drill down into one process, then you will see — as shown in figure 28.1 above — that you are navigating in one "slice" of the pyramid. For the Finance process, you would then see objects like "Accounts Payable", "Accounts Receivable", or "Asset Management".
  In our example, you would now see the map of one **state** that shows the major cities (*Main Process*).
- On level three, you will then see objects like "Create invoice" for the Accounts Receivable L2 process. Until now, you typically see processes

represented by value chain objects (chevrons) and these might or might not be connected to each other, depending on the data/meta model, and how your tool is configured.

This level would now be a map of a **county** that shows the major streets in the county and cities in our analogy. This *Business Process* typically shows the value creation on a higher level, and includes the process interfaces to other main processes.

- When it comes to the lowest level, typically the notation changes (to BPMN, EPC, or whatever other low-level notation your tool provides). This is the who-does-what-when-where level, and it includes tasks like "submit invoice" which typically have relationships to apps, data, roles, etc.

  In our analogy, this level would be the **city street** level map, where you can see the individual houses, intersections, etc. This model is the "*Process.*"

  What I typically don't recommend to model is the *procedure* level (aka *work instruction* level) below level four (in our example) – the level where you show which fields to fill out on a screen and which buttons to press. That information changes relatively frequently with every update and if you want to provide this information to your model viewers, I recommend taking a screenshot with callouts and embed it as an attribute or link to a screen recording by using the "link" attribute of a task object.

When creating this hierarchy, it is recommended that you restrain yourself to 4 or 5 levels. Any levels beyond that will just lead to an increase in the width of the pyramid, and therefore come with many, many more models to be managed.

What you also might see is resistance when agreeing on the number of levels because some org units might already have made up their mind and don't want to adjust their structure. Avoid giving in to this notion, so that your abstraction levels are the same. What I have seen in the past was that these types of org units would rather not have the "fight" with their stakeholders and all these middle layers (one client org unit had 12 levels, while we agreed with others on a 4-tier hierarchy) contained only one or two process chevrons ... this will severely impact your analysis, training, and reporting capability in the future.

How do I approach the creation of the functional process view?

Even though it would be beneficial to have a fully built out functional process view available from the start, I do not recommend spending months or years to capture all processes. Most likely, you will find plenty of things not being covered because people have left the organization, or existing material cannot be found anymore.

In addition to this, the time delay before you get to do something new is too long for the value that you get from that initial Herculean effort.

My recommended approach to this is as follows:

- Start with the **definition of the first two levels** (assuming a four-level hierarchy for the purpose of this article), and bring the stakeholders and dedicated process owners together. The objective is to define the *"buckets" of processes* and who takes over *responsibility* for which area.
  While doing this establish the *principle* that "he who owns a process, must also cover other processes requirements", which means, for example, that some lines of business are doing financial processes that are not done in the accounting department. If the accounting department is now tasked with building out the Finance main process, then it also has to model the financial processes that the line of business organization does – you should avoid to "outsource" this task to the line of business org unit (aka the other main process) because that won't break the silos, and you cannot ensure that whatever Finance activities they do will be consistent with the overall concept how Finance should be done in the enterprise.
  In addition to this, determining responsibilities and owners will also *clarify whom to involve* when it comes to creating End-to-End (E2E)

processes.

- After you've agreed on the high-level structure (the "buckets") determine the **concepts and principles for your main process**, and put these into writing. Also include **your plan of going forward** (which sub-area has priority for you?) and socialize this with the other main processes to get their buy-in, or trigger a change in your plans if they have requirements that justify a higher priority than your initial ideas.
- Next, take a look at two aspects: determine and collect which **content for your area exists** (and don't forget to ask the other owners if they have something to share) so that you can import lower-level processes into your repository as an accelerator.
- If you are using an enterprise-wide system, check if it comes with **reference processes**. Even though you might have done customizations, you should capture those (with a big sticker "needs adaption" on the model). The reference processes can also help to determine the shape of the content on the lowest value-chain level, in our case level 3.
- Investigate which **projects** are currently going on, and which ones are planned. Most likely, these projects have already created future-state content as part of their solution design, and you should add this to your catalog of existing content at this time.
  Use these projects to coordinate with them to ensure that they follow your concept, and once they have a final design — the "as-implemented" state, not what was signed off at the end of the solution design phase — grab this content and integrate it into your main process structure. Just keep in mind that the projects most likely created E2E processes and their share of your process might not be a complete process, so be prepared that you might have to add lower-level content to make a process "whole". We will cover how to keep E2Es and functional processes in sync further below in this chapter.
- While you are doing these lower-level activities (on L4 in our example), **adjust your level 3** models accordingly – the *Business Process* level. Depending on your conventions, which were defined in your technical governance phase of the tool implementation, you might show process

interfaces to other main processes (recommended) and therefore you will have to have *conversations* with them to determine the touchpoints and which data will be handed over.

You also will see that after a certain amount of content was created/imported on the lowest level, that you might have to *rethink the structure on level 3* – you might have to split or combine processes, or maybe an exception process that one of the Subject-Matter Experts has "sold" as its process is in all reality just another loop in another process.

What are the benefits of a functional business process view?

The benefits of a functional decomposition are multiple:

- **Consistent** leveling of processes, using a consistent meta model describing the abstraction level, and type of information that can be found in each level. Examples would be that you find applications on the lower-level processes (level 4 in figure 28.1 above), while risks could be found on either level 3 or 4 ... or both in some cases.
- This allows people in your organization to "**speak the same language**" when it comes to processes.
- Having a functional process view also allows ensuring that the processes in one slice of the pyramid **follow the same concepts and ideas**. In the Finance area, these might be that accounting only happens via cost centers and projects, but not via any other construct.
- When modeling your processes you will see the need to **coordinate with other areas** and determine what the **process interface** between two processes should look like: which end event in an upstream process triggers the downstream process, which data will be sent over, how applications support this, and so on.
- This exercise will also uncover the **various scenarios** that you might want to look at when it comes to the End-to-End view described below.
- Lastly, over time you will build up a network of processes that shows all activities of the enterprise **in one view**, and the responsibilities for each

area (on every level) can be clearly defined. This network can be used as the foundation for analysis of your Enterprise Baseline that you do when identifying improvement opportunities in the first phase of the process/solution lifecycle.

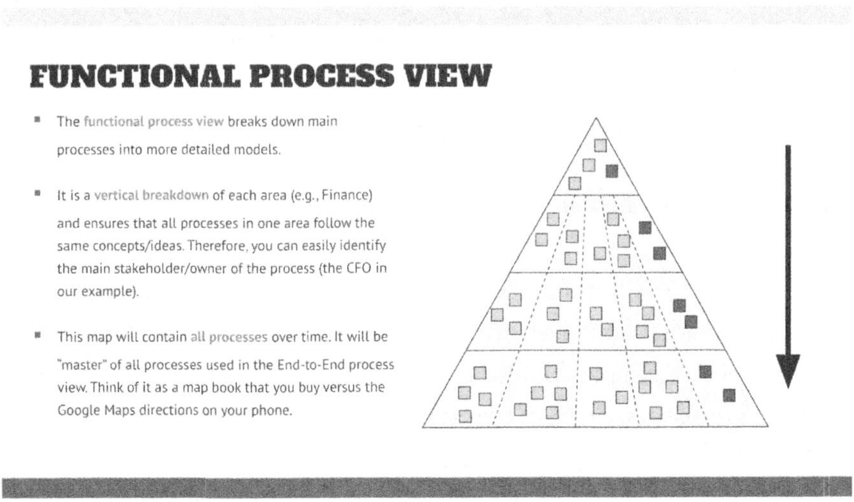

Fig. 28.2: Functional process view

End-to-End (E2E) business process view

A second process view that most clients build is the End-to-End view. These are processes that are assembled by processes from the functional decomposition (ideally variants of it, but we'll discuss this later). These are typically easily identifiable because they follow a "From/To" naming structure like "Order to Cash", "Procure to Pay", or "Hire to Fire". They are also defined by the start and end events, which are the bookmarks of the process, and typically these also can be used to define what is *not* included in the E2E process.

In addition to the start/end events, one of the first things you typically look at when defining an E2E is what the outcome of the process is, and how you can measure the success of it (using Key Performance Indicators mapped to processes and assigned KPI models that decompose those).

The main use case for E2Es are system implementations or other transformation projects, and process improvement initiatives. Typically, you see E2Es created in the **Solution Design** phase of the process/solution lifecycle.

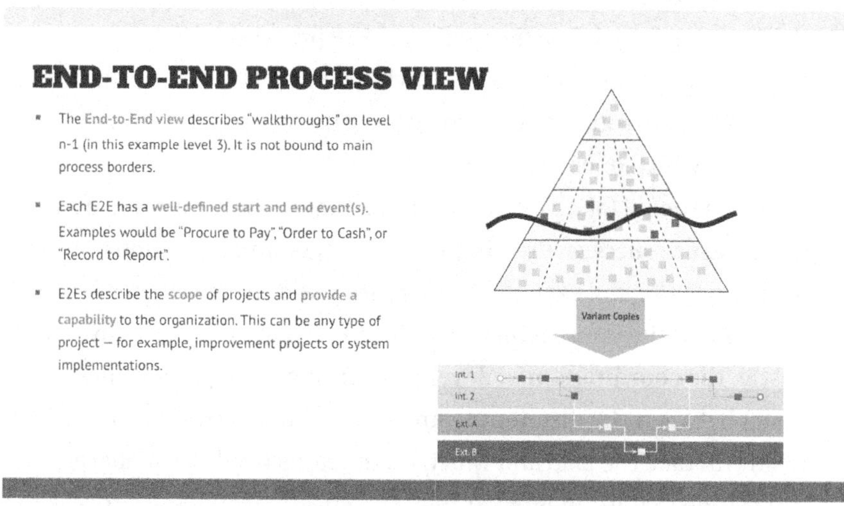

Fig. 28.3: End-to-End process view

To create an E2E, a bit of imagination is needed, and the first question is where to start and where to end. When making this decision, having a clear charter is helpful for obvious reasons – may it be the implementation of a new software ("let's update our SAP Finance module, it'll be fun" – nobody has ever said), or an organizational change.

In both cases, the E2E is describing the process part of a capability, and you should also check with your capability map which capabilities/capability areas you are tackling with this process.

How to create an E2E process

Creating an E2E process is a process in itself that includes not only your project team, but ideally other stakeholders and process owners as well. It also might be triggered by the needs of your project, so the PMO might be able to deliver information about the "why" and how the project is aligned to the company's strategy.

- The good news is, that you have already defined process ownership, so the first step is to define the **catalog of E2E processes**, to make sure that there is an agreement not only which E2Es shall be created and where they start and end, but also which priority in a program they should have and which ones are to be created first.
- Then you should take a look at the **performance indicators** and **how you will measure success**. Creating a baseline (pun intended) and putting it on paper is a useful exercise for measuring the effectiveness of the E2E process and the improvement that the project is tasked to deliver later.
- Lastly, the **coordination of development** (meetings, milestones, timelines) between the different teams should be determined, and which team will take the lead and which other teams need to collaborate and bring their requirements, so that all aspects of the E2E process are covered.
- Sometimes this can evolve into a situation of negotiations like in a Turkish bazaar, and it is recommended to **put the agreements onto paper** (a release plan), so that everyone involved and new projects have a foundation to look at when requirements or timelines change.
- After this overall agreement is reached, the modeling work can begin. The first activity here is to **identify the functional processes** on the lowest value-chain level (level 3 in our example, the "*Business Processes*") and to determine if they are fully or just partially included in the E2E. In figure 28.3 above, these are highlighted in blue.
- Depending on your database setup and the structure that you have agreed on in your meta model, you might have to **create higher-level**

models in the E2E Process View as well. Your E2E models should be on the same methodological level as the Business Processes in the Functional Process View.

- Then you can **create the E2E model** (value-chain) that includes *variants* of the Business Processes. Have a consistent naming convention for those in place, like "Variant of name_of_business_process – context". An example would be "Variant of Create Purchase Order – Material" [vs. Services that might be covered in another E2E]. Alternatively, you can create an attribute that indicates that this object is a variant copy of the (functional) master object and a small graphic can be shown at the symbol to indicate this.
  I recommend creating swimlane diagrams for those in which each main process has its lane ("Int 1" and "Int 2" in figure 28.3 above) and you also create swimlanes for external participants. Even though you will not model the content of those chevrons, you should include them so that it is clear what comes from whom, when process interfaces are defined (an example of an external process could be "Deliver material – vendor"). You also might think about adding an "external" boolean attribute, and having the external chevrons in a different color so that they are easily identifiable.
- For each variant chevron, now **create an assigned lower-level process model** (BPMN or different notation as defined in your conventions), just as you would create any other model. Add all necessary information that is required, such as apps, data, risks, transaction codes, etc. You also want to have a conversation with the process owners and see if they have already identified and imported existing processes, which then should be reused as an accelerator.
  Also model the process interfaces to the other variants in this specific E2E, and don't worry about the functional view for now. However, make sure that your start and end events are specific enough so that you can identify the process and avoid having a generic name like "Order created" (which order? A purchase order – and if so, for material or services? — or a task order?).

- The last step now is to **add the KPIs** that were defined in the initial definition/ planning steps, to either the level 3 E2E process and/or to the level 4 lower-level process. This information of what and where to measure the KPIs will be invaluable for others who will build dashboards or do process analysis/mining on the process in the future.

How do I manage my ever-growing E2Es?

Now that we have created an E2E model with its assigned lower-level processes, we are still not completely done. In larger projects, you easily run into the situation that you have defined your E2Es too narrow or too wide, or you realize that other E2Es also cover a process that is part of your E2E.

For that reason, a coordination is needed. Take a look at figure 28.4 below that shows three E2Es that are related to each other.

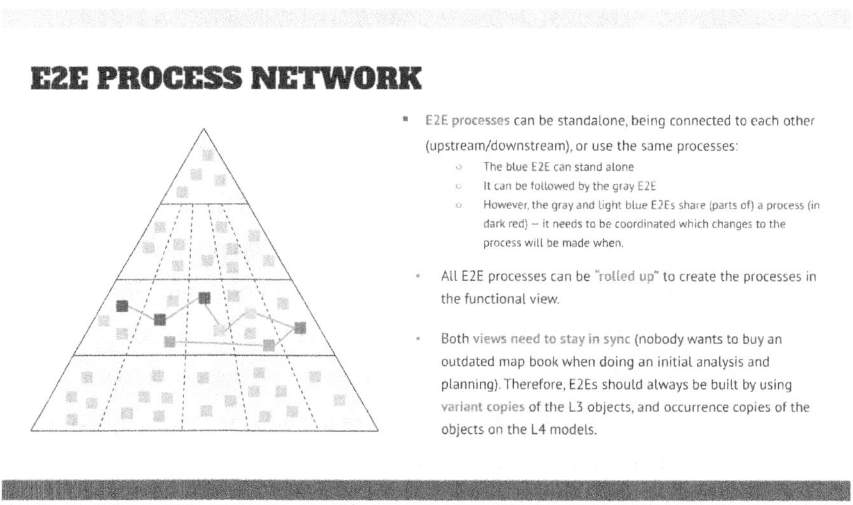

Fig. 28.4: End-to-End process network

- The *dark blue* End-to-End process can stand on its own. It has a clear start and end event and can be implemented.

- The *gray* E2E is a successor of the dark blue E2E process. So there needs to be a coordination with the dark blue E2E team that the process interfaces are well-defined – what is the shared event? Is it specific enough? Does the gray E2E require another start event that should have been created in the blue E2E (which means it was not really ready for implementation and needs to be updated – either now or in a next iteration of the blue process)?
- We also see that the gray and the *light blue* E2E process share a L3 process (indicated in *dark red*). Now you have to check if they share steps in the L4 process or if they target only specific steps that are independent of each other. If they share the same steps, then an agreement on those is needed, or a plan for a future development needs to be put in place.
- There is also the possibility of having an E2E that is fully contained in one main process ("Onboarding", for example, might only happen within the HR main process), which can confuse your users and needs a special step in explaining why this is a fringe use case, and why you don't want a "From/To" process name in your functional process view. If you already have a lower-level process in the functional view, then I recommend creating an attribute "Implementation Status" for these functions where you can select "in planning", "in implementation", and "implemented" and have them highlighted using your template so that it is easily identifiable which steps of the functional process still need to be implemented when doing the next E2E planning exercise.

However, the only way to really manage all E2E processes is to plan their development in releases, and have a predictable schedule that not only avoids that someone misses that they should participate in conversations and decisions, but it also allows communicating what is "up and coming" to your end users/viewers.

How do I keep both views in sync?

As indicated above, it is crucial that you do not see the E2E view as a standalone process view. It is tied to the functional process view, and is reusing the objects on the lower level, and managing dependencies and variations by using variants on the value-chain level.

Figure 28.5 below shows the activities in a release as overview:

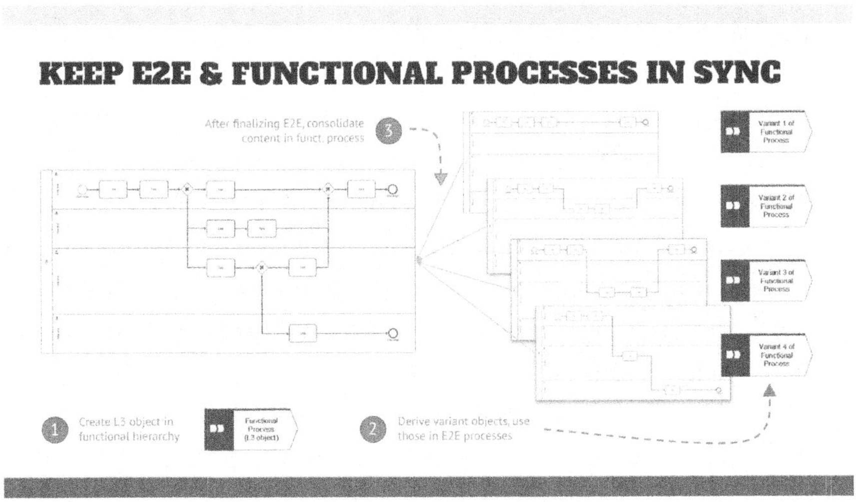

Fig. 28.5: Approach to keep both process views in sync

1. **Identification of a L3 process** in the functional process view, represented by a chevron.
2. **Creation of variants of that object** in the E2E process view and assembly of L3 E2Es using these variants. Thereafter, **new lower-level models** can be created that are assigned to the variant objects. These use occurrence copies of the L4 functional process (if one exists), and will be modeled as needed for the E2E process (gateways, process interfaces, etc.).

3. Once the lower level E2Es are finalized and approved, the content of those can be **copied into the L4 functional model** (the left diagram in figure 28.5 above) and it can be approved, versioned, and published. Keep in mind that you might have to adjust the model and adapt it so that it fits into the lower level of your functional process view.

In my experience, the coordination effort for the releases is minimal in the bigger scheme of things and well worth spending the time for coordination.

What are the benefits of E2E processes?

There are multiple benefits of having End-to-End process models:

- The E2Es describe the **functional scope** of a project ("what needs to be done"), for example, in a business blueprint design for a system implementation.
- They can also describe a **capability** and make it "real" for non-architect types of users. After a successful implementation of an E2E process, the organization has gained a capability – it "can do things" in this configuration of process, apps, data, roles, etc.
- They also are the **basis for process improvement** efforts. The KPIs that are defined can be measured in either traditional process analysis methods, or using data-driven analysis such as process mining. Unfortunately, the KPIs that are delivered with reference_content, such as the APQC Process Classification Framework, are aligned to the functional process structure, and you will have to do some re-shuffling when aligning them to your E2E processes if you want to use them as an accelerator.
- Lastly, E2E process models will **complete the functional process view** (the "big map") that is needed to have an overview of "what runs" for analysis in the Enterprise Baseline.

But what about risks?

I have to admit that I worked in projects that were mostly driven by Governance, Risk & Controls (GRC) folks. Even though they can be a powerful ally – think about the amount of fines an organization has to pay if they are not in compliance, which can help to make the business case for implementing a process management capability easy – but they also have a different "view of the world". It is not driven by looking at how an organization performs, or how it can improve its processes, but with the lens of risk identification, avoidance, and mitigation.

This can lead to the prime focus of managing the risk inventory and aligning them to processes, independent of which level that process exists, or if they are even related in a way that makes sense for the business. They also might add processes because certain apps or data are used in them that are related to the same class of risks. Overall, it might create a picture as shown in figure 28.6 below, that neither fits into the functional decomposition, nor into the E2E structure ... both views visualizing and structuring how the business "works".

In one of the projects mentioned above, it even went so far that we created a "risk-oriented process view" that stood on its own, and combined processes based on risk-related criteria. One example was a high-level process of "Account creation", which was on the one hand side not very specific, and on the other hand it did not distinguish that the creation of an account for a credit card might be something entirely different from creating an account for a mortgage (or an IT account for online banking).

Needless to say that the org units who were not involved in setting up that initial structure, but were supposed to add their processes to it as "fast followers", could not work with it because they didn't recognize themselves in this structure. This brought the project close to a failure after the successful initial push of appeasing the GRC requirements.

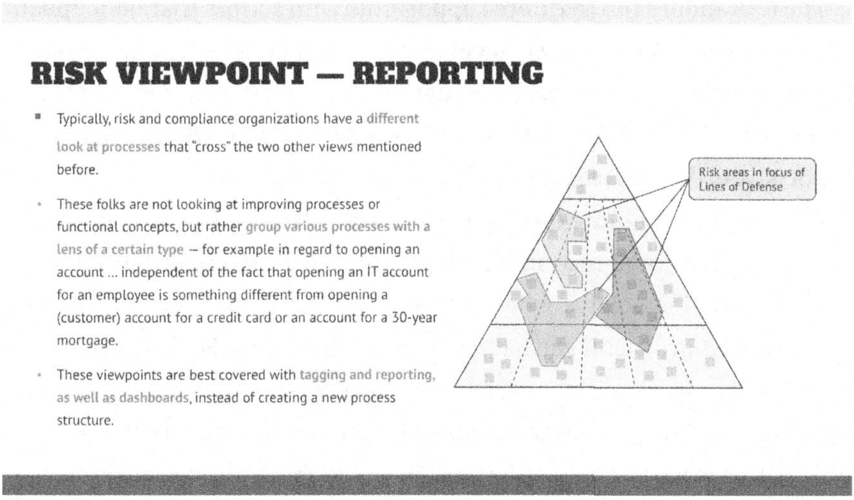

Fig. 28.6: Risk and Compliance viewpoint

But how do you treat the legitimate demands of the GRC folks in all three Lines of Defense (operational risk management, the risk system, and internal audit) then?

My suggestion is to treat this topic not as a separate view, but to **make it a reporting exercise** that will provide the necessary reports and dashboards for the GRC stakeholders. This can be done by tagging the steps where risks occur and then modeling which controls mitigate these risk and how they will be tested (and who does the management of risks, controls, and tests). This structure, including the attributes of risks and controls, such as type, frequency of testing, etc., can be sent to a GRC tool which can automatically create audit plans and other things that are needed to manage the GRC requirements.

An architecture tool should also support the **creation/import of existing catalogs** for higher–level laws and regulations, regulatory obligations and requirements, business rules, policies, risks, controls, etc. so that business users who create process models can pick from a list and don't have to create their own objects that then would need to be reconciled "after the fact" by admins and GRC folks.

The tool should furthermore support this with scripts that help **"push" higher-level risks to lower-level steps** and to create the **relationships between tasks and controls** (and thereby reducing the amount of manual work). Other useful features would be **predefined workflows, scheduled reports, and dashboards** for the GRC stakeholders.

In general, I would expect the risks and controls to show up in the functional view, and recommend handling the development in the same ways as described above when we were talking about how to keep both process views in sync: If you have risk and controls already identified in your functional processes then you are good. However, you should invite GRC folks to your E2E process development sessions, and model their input as well, so that when it comes to the third sync step above, the new relationships between tasks and risks will make their way into the functional process view for future analysis and auditing.

Conclusion

Both process views go hand-in-hand with each other when it comes to defining and describing your business processes. Having just one of those is not sufficient when you set up a full architecture in your tool. They will also help when it comes to coordination with the risk folks in your organization, which can be – depending on your relationship with them – your biggest friend or foe.

29

Import Of Existing Content - A Blessing Or A Recipe For Chaos?

One major point of the implementation of an architecture tool is the import of existing content. Organizations have years and sometimes decades of content in various formats available, that might be in various statuses – from "current" to "horribly incorrect", "cannot be found anymore", or "outdated".

However, it might be a worthwhile exercise to get your new program started by bringing in content that your users and stakeholders already know. This will not only add meaningful content and might act as an accelerator for future projects, but also creates trust in your new repository when users interact with it for the first time.

In this chapter, I will walk you through an approach that has been proven to be successful in multiple import projects that included dozens and hundreds of models and other input documents. In the "how-to / import content" part I will focus on the import of models (in Visio, *.bpmn, ArchiMate,[67] or other formats).

Content import overview

Importing content is straightforward, but also a process with many moving parts in the air. Therefore, it is important to not lose track of where each individual model is in its journey.

Figure 29.1 below shows the process in an overview, and it assumes that you already have a database structure in place that includes a hierarchy of items. In this chapter, I am using the import of process models (BPMN models to be more precise) as an example, and show steps aligned to the architecture tool that I am using throughout this book – ARIS. Depending on your chosen tool, some steps might be slightly different.

In addition, please note, that I am only covering the import of existing models, not the import from Excel, a database, or another source system using the tools' APIs. The steps are similar, but it would take significant more details to explain those, and would "muddy the waters" of the "import an existing model" use case.

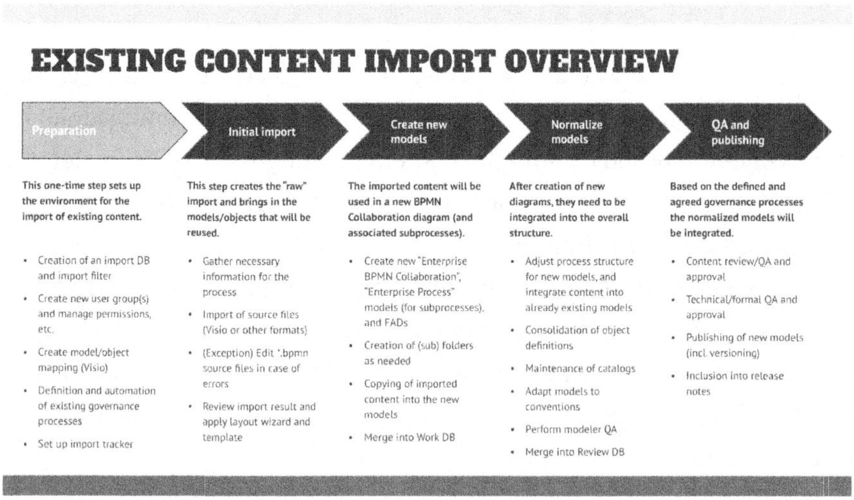

Fig. 29.1: Existing content import in an overview

The process consists of five steps:

- **Preparation**: This is a one-time step that sets up your environment for the import of the existing content.
- **Initial import**: This step brings in the "raw" import and brings in the models/objects that will be reused.
- **Creation of new models**: This might be a BPMN-specific step because of the implementation of BPMN in different tools and the export "flavor" of the source system.
- **Normalization**: In this step the imported models will be integrated into predefined structures, and catalogs will be maintained.
- **QA and Publishing**: The last step covers the various quality assurance and approval steps to finalize the import and make the content available for end users.

Phase 1 - Preparation

The first exercise is to **create a catalog** of to-be imported content. Ideally, you do this in a database that will be the "source of truth" through the complete import project, and also supports the estimation of efforts that will help with prioritization of the import sequence, or determine which models will be imported and which ones not (we are big fans of Notion[68] over here at What's Your Baseline, so you might want to take a look at this tool).

When creating this overview, I recommend involving as many people as you can – stakeholders, business unit leaders, project managers, architects, and start with a loose collection of ideas and content. This shall be oriented to the goals of your organization (and yours, as defined in the Strategy phase of your implementation), and the objective is to get a rough overview of "what is out there" and what the volume of this content is.

Your stakeholders can be used as multiplicators/communicators and can open doors to people in your organization that you had no idea that they existed, or that they have worked on projects that are relevant to your effort.

Once you've created a rough overview, align the models to your priorities and initiatives, and determine an initial prioritization and sequence for import. Discuss this initial version with your stakeholders and get an agreement on which content shall be imported first.

From what I've seen in the past, everything that is regulations/SOX-related is a viable candidate, as well as existing taxonomies in the various architecture views (not only in processes, even though the examples in this article will focus on process models).

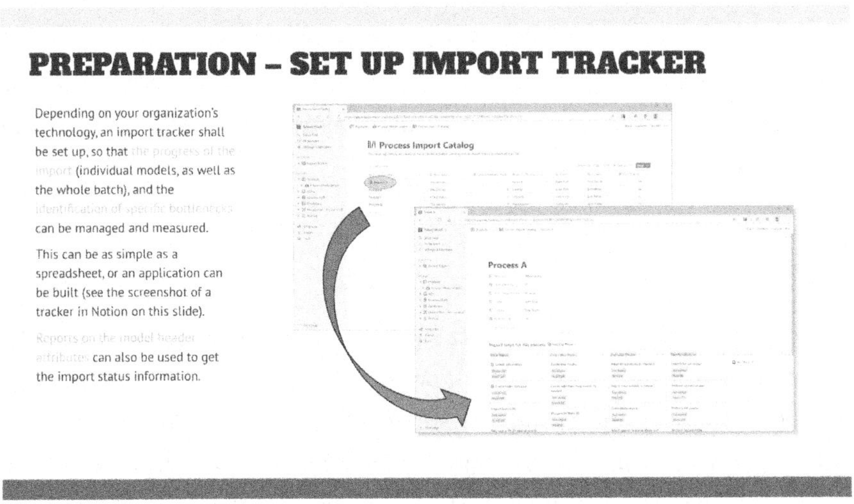

Fig. 29.2: Preparation of the import tracker

After you have created the initial overview, it is time to discover the details of your to-be-imported content. This includes descriptions, complexity, source file format, hierarchy level, (process) area, owner, (modeler) contact, assigned modeler (importer), and the planned import batch.

Especially the complexity will be important to estimate the effort. My recommendation is to keep this simple and just count the number of objects that need to be imported, and then come up with a simple rule that defines low/medium/high complexity (20/50/50+ is a good start). Moreover, keep in mind that the source file might include multiple models (tabs in Visios for example), and that those are counted as individual models.

The next step is to **set up your database structure and create an import filter**. As shown in figure 29.3 below, you want a separate (logical) database to do the initial/"raw" import of the source files. Often, these include content that you do not want to have in your governed DB setup (in the gray box below), and you also might need elevated permissions or a special method filter for the import that is different from the standard filter you are using in your day-to-day work.

An example for this might be that you are using the "Enterprise" BPMN model types for your regular modeling because you want to take advantage of the ability to use dedicated org/role/app lanes in your collaboration diagrams. When importing *.bpmn files from another system, these are not known in the technical file format (because the specification does not specify these, but is rather "foggy" in describing what lanes are). Therefore, ARIS will import all *.bpmn files as standard BPMN Process Diagrams. The lanes in those won't be reused in the later "create new models" step and therefore should not go into the Work DB where the normalization happens.

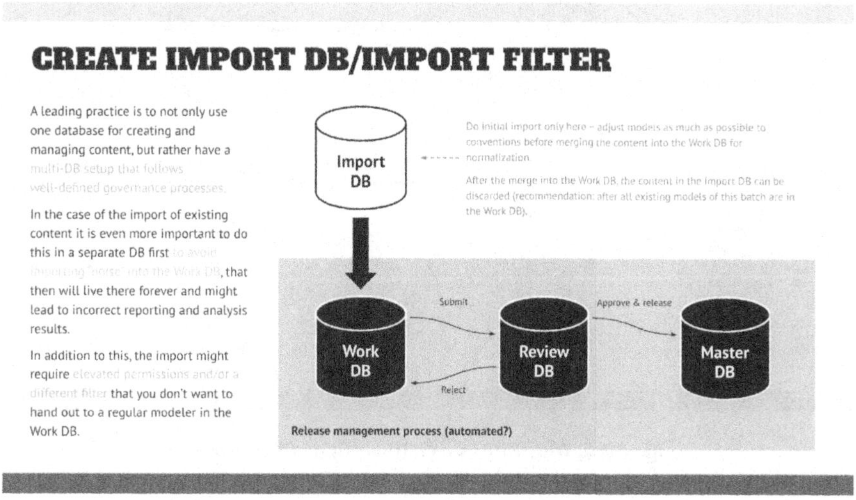

Fig. 29.3: Preparation of import DB and import filter

Once you have merged the content into the Work DB — before the normalization step (see part 2 of this series) — the models in the Import DB can be discarded. A good practice is to have a "clean" Import DB before the next batch of models is ready for import.

Next, you might have to **create a mapping** if the source files require it (Visio for example). This step is not required for every source file type (e.g., BPMN, ArchiMate, DMN) because they have a technical format defined, and tool vendors take this and create out-of-the-box imports for these formats.

In the case of Visio, you would have to map the Visio stencil to a model type, map the shapes to object types, and map the attributes of the source shapes to model and object attributes. Obviously, the more consistent things were created in Visio, the less effort is needed for the mapping – if things are all over the place, you might want to ask if the better way would be to use the Visios as an input for a manual creation because the preparation of each import, and the cleanup after the initial input might be more than a manual recreation.

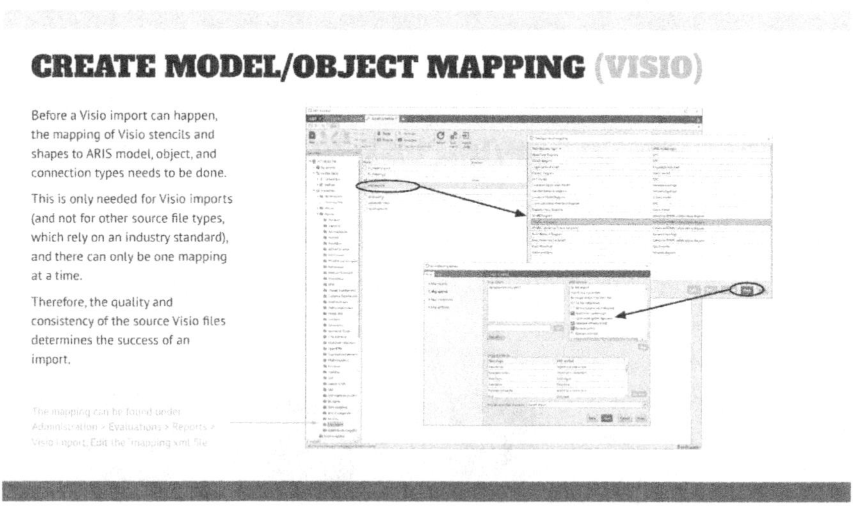

Fig. 29.4: Mapping of Visio stencils and shapes

The last step in this phase is to **determine and implement automation opportunities**. Some architecture tools allow you to create workflows, or other automation means (scripts, macros, etc.) to trigger activities that can make things easier in the overall process. The obvious use case is the release cycle management shown in the gray box of figure 29.3 above, but there are also smaller automation opportunities, such as merging of content between databases, triggering semantic checks, or locking/unlocking models if they reach a certain status in the import process.

Phase 2 - Initial import

In this phase, the "raw" import will be done, and this is the step that you will see in the vendor demos when the topic of import comes up.

The steps in this phase are:

- **Gather information** – besides getting the source file that is to be imported, try to get as much information about the process as you can get. This can be related documents (such as risk documentation, process manuals, SOPs, or policy documents) and -as a minimum- a graphic of the existing model, so that you can lay out the imported content for the initial review accordingly.
  Also get the contact information of the modeler of the existing model, and their availability. And lastly, inquire about the age of the content – it might be horribly outdated, and that could take this model off the list.
- **Create folder structure** – in the Import DB, I recommend creating a folder for each to-be-imported model, and then create two subfolders: "raw import" and "edited model". The first one will be used for the technical import, while the second will be used in the next phase.
- **Import source file** – in this step, you perform the import steps as needed for your tool. In the case of a BPMN import, unzip the *. bpmn file and look for the "main" process (there could be many in that file, and also *.bpmn files that you don't need, such as "resources" or "glossary"). This main process will become a BPMN Collaboration

diagram later, while the others will be BPMN Process diagrams, as they are subprocesses.
Import each individual file from the Zip into their folder.

- **Edit source file** – depending on your architecture tool, it might perform a check on the to-be-imported file, and give you a warning (what might not be importable), or an error in case the code is not correct. In the latter case, it most likely will tell you on which line the error resides – open the *.bpmn file in your text editor, and delete the corresponding tag (from <xyz> to </xyz>) and try the import again.
In most cases, I was successful with this, and the reason for that behavior is that the various tools have their interpretation of the technical spec, and add custom tags to the code that just their tool understand (and needs). Typically, this does not create any unwanted issues, but be prepared that you might have to do this a couple of times for a file, or you end up in a dependency nightmare – in these cases it might be easier to simply create the model in your architecture tool "by hand",
- **Review the imported file and do some initial formatting** – the tools are getting better in their imports, but occasionally the X and Y coordinate information might be lost, and you will receive an imported model as shown below. In these cases, I recommend doing these three steps:

1. Select all objects and set their representation to the tool's default (e.g., size of objects).
2. Apply the template for visual representation of your model.
3. Use the layout wizard to let the tool determine the best layout that the algorithm can do.

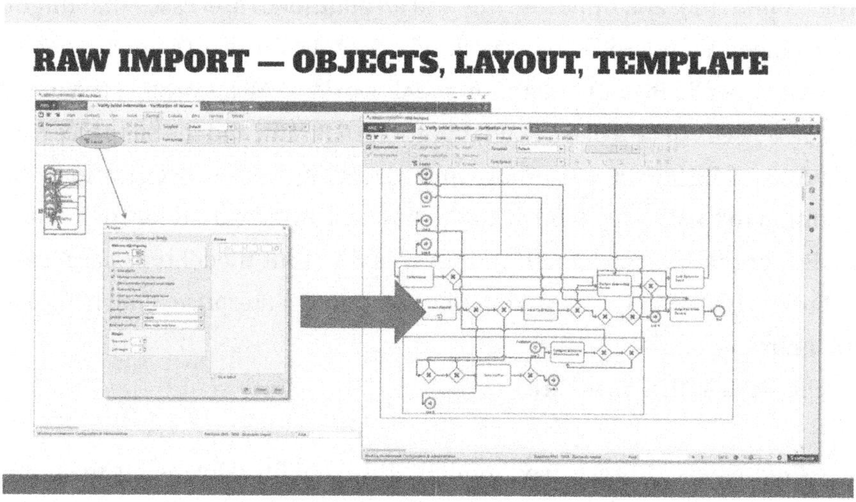

Fig. 29.5: Source content after raw import

After you've imported the models, you might want to do additional "pixel pushing" to give the model a more user-friendly layout, and also correct immediate errors that might stick out. In the screenshot above, for example, all events were imported as either intermediate or end events, but no start events. Go through your model, and change the symbols of the event objects as needed (the graphic from the source file will help you determine this).

Up to this point, we accomplished what the tool vendors show as "easy-peasy" content import, and their claims that the tools can import source content is correct.

Phase 3 - Create new models

This phase is most likely specific for our example here because we are looking at BPMN process models and the limitations of the specification,[69] which in some areas does a not-so-stellar job in specifying things (for example when it comes to the use of lanes).

Up to now, we have worked in the "raw import" folder for your models in the Import DB. We have identified the main process that is captured in the *.bpmn file, and imported it – and all additional processes that might be included as subprocesses – as standard BPMN Process Diagrams.

This typically is not enough for a real-world BPMN project – the main process needs to be a BPMN Collaboration Diagram, and the subprocesses need to be BPMN Process Diagrams according to the specification. In addition to this, most clients want to have a semantic used for their lanes (roles, org units, apps, etc.) and for this different model types are used ("Enterprise BPMN Collaboration Diagram" and "Enterprise BPMN Process Diagram").

The steps in this phase are:

- **Create new "Enterprise" models**: this step literally means to create the correct Enterprise BPMN model (type), and copy the objects from the raw import into them. Then compare this with the graphic and add missing objects that might not have been imported, or were of the wrong type, to the model.
- Also capture (in a separate file or your tracker) incorrect modeling that you had to correct (some tools, for example, allow stretching subprocess objects over multiple lanes which is not correct according to the spec), and maintain model attributes, such as name, description, status, contact, etc.
- **Create subfolders and models as needed** – repeat the previous step for each subprocess. This will create a correct folder structure in your tool's Explorer view and will make finding and moving catalog items into their correct locations easier in the Work DB later.
- **Merge into Work DB**. Now that you have created the model as good as possible in the Import DB, it is time to merge the result from the "edited model" to the Work DB. The procedure for this might be different in your tool, but I recommend creating a high-level "import" folder in your Import DB, and moving the new models *with their objects* into that high-level folder.
  This will avoid that objects being stored all over the place once the merge into the Work DB has been done. You might have to coordinate with

your administrator on this step in case you do not have the "Database Management" function privilege to do the actual merge.

After the merge is complete, the models in the Import DB are not relevant anymore and can be deleted to create a blank slate for the next batch of imports.

Phase 4 - Normalization

This phase drives the effort in a model import project – it will integrate the individual models into the existing structure/hierarchy of your Work database and is a mix of conversations, model adjustments, and admin tasks.

Before you begin to normalize your models, I suggest that you do an **update from upstream systems** (in case you have created interfaces to other "systems of truth" as part of your architecture tool implementation). This will ensure that you have the latest and greatest objects in your catalogs, and you don't run into the situation that you see gaps, which might lead to delays if you have to contact the upstream system owners to ask for what the correct object will be – for example applications, or risks.

Typically, these updates are scheduled (e.g., once a week), but I recommend asking for an extra update and wait for the notice from your admin that the import was successful.

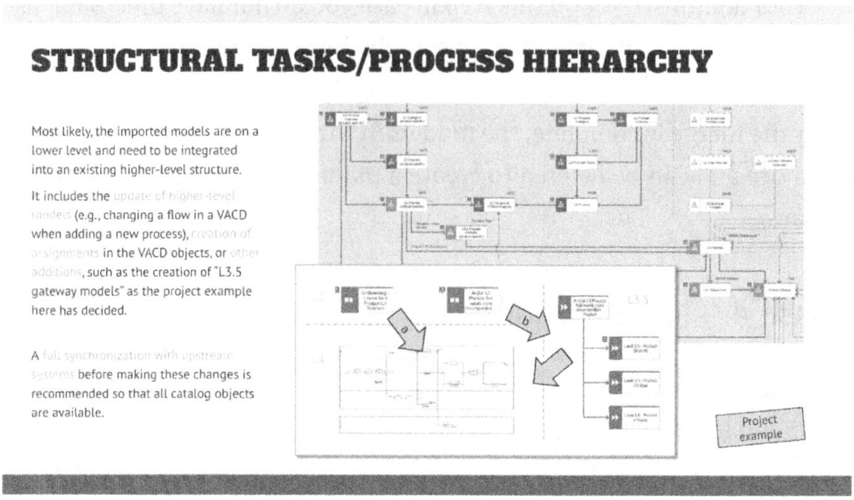

Fig. 29.6: Structural tasks during normalization

After the update from upstream systems, the steps in this phase are as follows:

- **Integrate your new model into the structure** and **adjust other models as needed** – this means a couple of things and opens a couple of questions.
- *Is your new model a "standalone" model* that already has a place identified, so that your BPMN Collaboration Diagram will be a full lower-level model? Or does it contain content that needs to be split over multiple existing models because it was an "End-to-End"(-ish) diagram that covered aspects of various, already existing models? In that case, the elements of your process need to be integrated into the already existing models, and there will be no separate new model.
  Before you do this, you might want to not only consult the other owners of the existing models in the Work DB, but also with your model contact from whom you have received the source file that was to be imported. Get agreement from all affected people and escalate if you cannot reach that agreement.
- *What about your subprocesses?* Are they really a subprocess of this

process, or do they already exist in the database? If so, are they also "main" models, which would mean that you will have to create process interfaces between these processes, or do they exist as subprocess models (BPMN Process Diagrams) elsewhere? In that case, you would have to update the BPMN Process Diagram and bring the end events "up" into the main lower-level collaboration diagram (remember: BPMN has a specific way how to do this correctly).

In every case, you will have to contact the owner/modeler of the other models and agree with them how to model things. For example, on a process interface the end event of an upstream process is the start event of the downstream process, and it should be the same object in your database. If you cannot agree on a specific name, the golden rule is that the upstream process determines the name of the end events.

- *Check if you have workarounds in your meta mode*l as shown in figure 29.6 above. Even if this is just a temporary convention, you might have to create higher-level models and then assign your new model to one of those objects (in the example above, the structure in case B was not granular enough and therefore a "Level 3.5" Function Tree had to be maintained that included the detailed variations of the L3 process).

  An example could be that the taxonomy just defined a process called "Create account" – now you have to define which types of accounts and the different ways to create this. This could lead to L3.5 objects named "Create Credit Card Account – Branch", "Create Credit Card Account – Online", "Create Credit Card Account – Telephone", "Create Mortgage Account – Branch", and "Create Mortgage Account – Online". Your new collaboration diagram would then be assigned to one of the five objects.
- Once you have integrated your new model into the correct place in the structure, you need to work with an admin to **consolidate objects**. Since your new model did not have an ID for each object that your architecture tool recognizes, new objects were created, which leads to the situation that you have multiple objects with the same name.

  To solve this situation, you need to reach out to the object owners and determine with them if you **really** mean the same thing, and it is an object, for example a process step or event, that shall be reused. Typically, that means that the step uses the same systems, data, is done by the same role, and has the same attributes (e.g., the definition) – if that is not the case, you have a new object, and its name should be different.

In case you agree that it is the same object, work with your admin to do the consolidation. Good architecture tools have a feature that does this in an admin view, which will then replace all objects that need to be replaced with the already existing object in the background so that there is no need to open each model and make the changes manually. This background step will also delete the to-be consolidated objects from the database, to avoid having "Zombie objects" in your database that would spoil a future analysis.

- The next step is to **adapt your model(s) to the project standards and conventions**. This might change the layout significantly from what you have received as model graphic from the source system, but typically those models were created "in the wild" without any constraints, and therefore might not follow any of the standards that are defined in your program. It might be a good idea to take some notes of what you've changed for the upcoming review with the owner of the source model, so that you can explain why you made the changes.

  Also note down things in your current conventions that forced you to do "interesting modeling constructs" when adapting the models, and submit those to the method owners as improvement suggestions for the next version of your standards and conventions.
- After you have adapted your model(s) in the Work DB, schedule a **meeting with the original model author/contact person**, and walk them through the new models. Capture their improvement ideas and get their buy-in so that you can close out the content modeling after the next iteration. Avoid having an unlimited number of iterations and reviews, but set the expectation that you have only one or two iterations. If your model contact does not agree with the normalized models after

the maximum number of iterations, escalate the situation – these iterations can be significant time drains, and might bring the whole import project into a risky state.

- The last step in this phase now is to do a **modeler QA** on your model. Your project should have a checklist, as shown in figure 29.7 below, that includes a mix of visual inspections ("is the happy path clearly visible?"), semantic checks ("was it modeled correctly according to the notation standards?"), and reports.
  You also might be triggered to maintain model attributes that will help with the QA and publishing activities in the next phase.

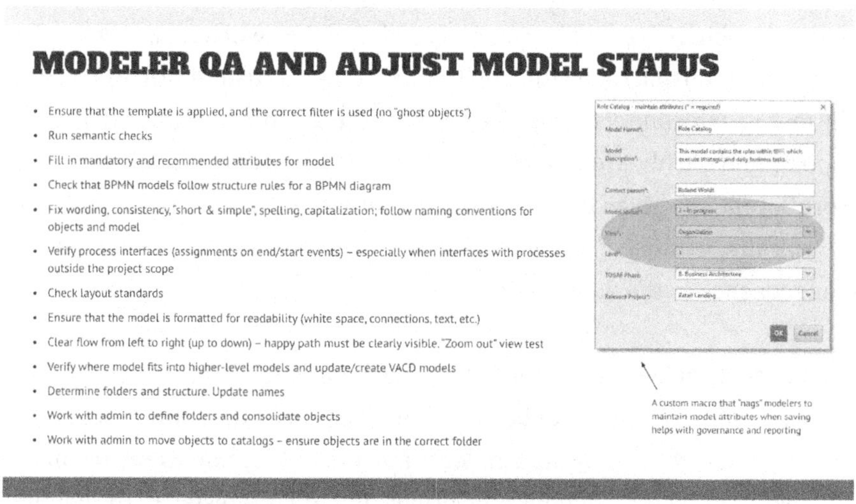

Fig. 29.7: Modeler QA checklist

Phase 5 – Quality assurance, publishing, and regular certification

The last phase of the model import contains the formal approval process and various merges, versioning and publishing.

The first step in this phase is the **submission to QA review**. Even though it sounds trivial, I suggest automating this whole phase with a workflow engine that your architecture tool should have. This will take out the minutiae of locking/unlocking the model in the Work DB, check of mandatory attributes, capturing of timestamps for reporting, and merge of a submitted model into the Review DB, where stakeholders will do their reviews.

Another benefit of using a workflow engine is that stakeholders will be notified when their contribution is needed, and they will have a task list that they can work from (and either do the task, using a custom GUI, or delegate the task to someone else if needed). Lastly, the workflow engine can provide reports and dashboards that show where the individual models are in the workflow at any given time – a feature that project management will love.

Typically, you see three types of review, and I recommend doing these in this sequence in the Review DB:

- **Content review** – this is done by a stakeholder from the line of business, who might or might not be the original author of the source model that was imported. These stakeholders are most likely not interested in the details of your modeling conventions, but are interested in the content correctness of the model.
  This might lead to conversation needs where you, as the modeler, might have to explain why things are modeled in a certain way, and why a convention exists.
  Use the collaboration features of your architecture tool (commenting and "@ tagging" for notifications) instead of rejecting the whole model in case you, as the reviewer, are not in agreement or do not understand what is presented to you. A rejection will "reverse" the

whole submission and force a new submission for approval. This will cost precious time.

- **Risk review** – this might be a step that will not apply to all models, and therefore you should have a checkbox on the submission for approval form that will trigger a risk review ("SOX-relevant" or similar).
  The risk reviewer will most likely act similar to the content reviewer, even though their focus of the review is different from the first review – e.g., "are the correct risks and controls included in the model?".
- **Technical QA** – the last step in the review is a formal QA, and is typically done by a central team that ensures that the standards and conventions are used. To accomplish this, these reviewers might use a similar checklist as was used for the modeler QA above, but in addition to the formal aspects, the focus here is to review the integration of the model into the structure (correct interface modeling on all levels, not only on the lower level, correct consolidation, usage – and potential consolidation – of catalog objects, etc.).

If any of these reviews fails, the process will be sent back to the modeler with a list of items for correction, and all technical steps in the workflow will be reversed (unlocking of model in Work DB, deletion of model in Review DB, notification of all people involved about rejection, etc.). The model then will be re-submitted after the corrections are done, and it needs to be defined in the workflow if all three review steps need to be done again, or if existing prior approvals are still valid and a later reviewer will be triggered.

In case of the successful approval, the **fully approved model will be versioned, the model status being set to "released" and it will be merged into the Master DB** for consumption by all viewers in the organization, as well into the Work DB, so that the next iteration of this model starts with the identical model that exists in the Master DB.

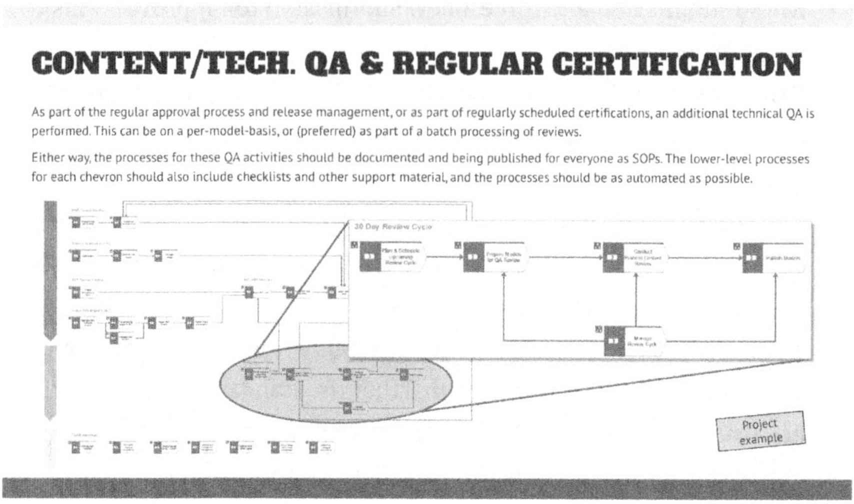

Fig. 29.8: QA process (content and technical QA)

As shown in figure 29.8 above, there might be another use case for the approval process that was explained in this section – the regular certification of models. Even though many steps are identical to the "regular" approval process, there are additional steps needed – for example the assembly of "review batches" of models, the management and reporting on the certification process, and the certification itself.

I recommend defining this as a separate governance process and implementing a separate workflow in your automation component of your architecture tool for this purpose. A detailed description of the certification process is out of scope for this article for that reason.

Model import conclusion

After going through all five phases of the model import process that were described in this chapter, you have successfully imported and integrated the new content, and also published it while getting approval from all stakeholders.

From now on, the imported processes will follow the established governance, and the source files should be deleted, so that no new changes will be done to them – a tempting idea for some people who don't like to "do the overhead" that might come with working in a central repository. I suggest that you discourage this practice and simply ignore if someone shows up with "their version" of an existing model ... easier said, than done (depending on whom that person is), but be assured that eventually you will be better off insisting on the consistency of the repository.

Bonus offer

A major takeaway from the process shown above is to keep track of all the moving parts when importing dozens or more models. A tool that is excellent at managing this (and other things) is Notion. I am a big fan of this tool for organizing and preparing our podcast, and it is affordable – free for personal use (one owner, and you can invite guests for edits) or working in a team.

The tracker covers not only all phases of the import process (catalog, tracking each model through the defined process in Kanban boards, and reporting), but also includes additional information on how to do this. It also includes a presentation with detailed step information and examples from a large import project.[70]

30

Base Process Analysis

Getting started with process analysis can be hard. There is nobody who can give you a hand, and the tools have documentation about this to varying degrees. However, there is a kind of "base" analysis that does not require dedicated tools (besides your architecture tool), or additional data collection, for example, for simulation or process mining. Those "advanced" analysis methods are discussed in the next chapters.

This chapter does not aim to be a complete guide to process analysis and all the different angles one could dissect a process and find improvement opportunities. It is meant to start the conversation about the need for "constantly improving" and how to do this. In addition to this, all the measures can be captured by hand, but a good architecture tool will have reports and other means available out-of-the-box so that you can avoid painful counting.

Why process analysis?

In the past, there were multiple waves of process improvement that aimed at different aspects, one approach (the Business Process Reengineering) aimed at rethinking the value-streams that are created by processes, while others (Lean/6 Sigma) approaches looked for smaller improvements in a given lower-level process.

In this chapter, we are focusing on a single process or a combination of processes that are assembled in an End-to-End fashion. The two terms that are guiding the base analysis — or any process analysis for that matter — are:

- **Value** – the measurable business outcome of a process. This can be the ultimate outcome for a customer ("We solved a problem", for example), or be a sub-outcome in a longer series of processes that ultimately will end in a customer outcome. An example of that could be the creation of a purchase order in a Procure-to-Pay process, which is needed to complete a first step of that process and trigger a vendor to send a product or perform a service.
- **Waste** – this includes all activities that do not contribute to value creation, but are performed "because we have always done it this way". The Lean folks have a very tight definition of waste[71] that I encourage you to take a look at. Examples of that are incomplete process execution, or a low quality of the outcome, which will require re-work that could have been avoided.
  Please note that this does not mean that things like documentation of systems or things required by regulators should be eliminated from your processes, just because an end customer does not get an immediate benefit from it. Those activities are value-creating, but they serve other stakeholders than a paying end customer – a future IT team, for example, that is tasked with updating a system. Or the public that can be assured that things are done properly within the boundaries of the regulations in the two examples above.

When analyzing a process, you always should have those two aspects of process improvement in the back of your mind and evaluate the metrics below on those two things.

When looking at the **solution lifecycle**, that we've discussed in an earlier chapter, base analysis typically happens in the first two of the five phases – in the High-Level Design phase or the Solution Design phase. There is no hard rule on which type of analysis (higher or lower-level processes) shall be done in which phase. However, the examples below focus on a lower-level process analysis.

The examples below also don't have the idea of being a complete set of metrics to capture and things to look at. But they are meant to give you a head start in case you have never done this type of work before. I encourage you to spend some time on additional research on this topic (maybe also by reaching out to your tool vendor or consulting partner) and develop your own set of standard metrics you want to analyze your processes against.

The important point is that you get started and define **a set of analysis that will be packaged as a product** for all of your base process analysis. This will include tools, such as spreadsheet templates shown below, that might do calculations for you, or additional material that you will use to train new people in your tool set.

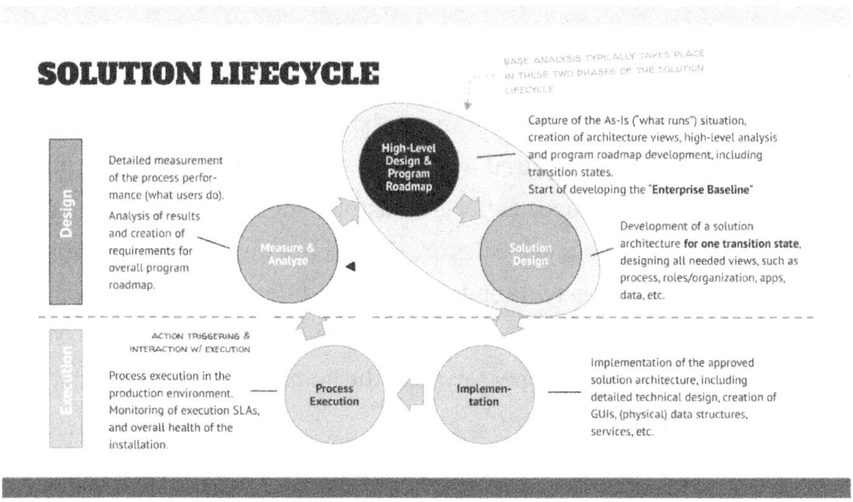

Fig. 30.1: Base process analysis in the solution lifecycle

What are common analysis areas?

Since the number of things you could look at when analyzing your process, I recommend to focus on these five major areas as part of your "base process analysis" product:

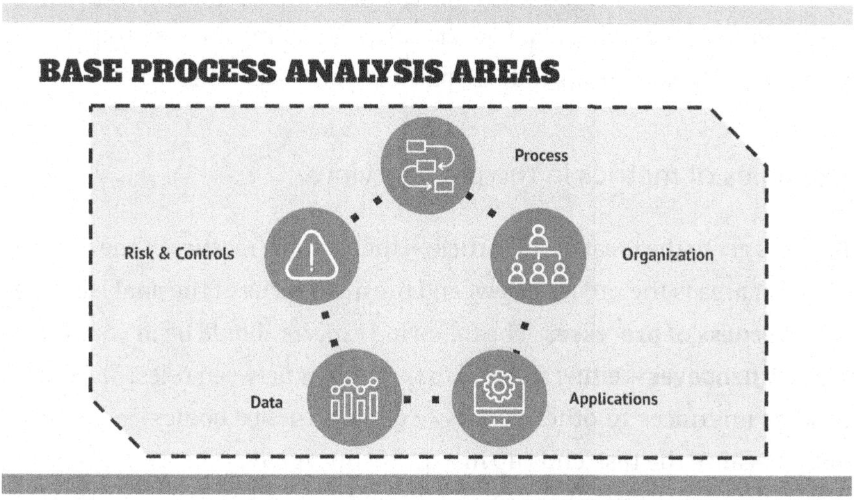

Fig. 30.2: Base process analysis areas

- **Process**: Number of processes/tasks, org changes, tool changes, internal and external handovers, approvals and checkpoints, pain points, process gaps.
- **Applications**: Number of tools, tool redundancies, unique tool combinations, technical interfaces, workflows/integration.
- **Organization**: Number of org units/roles, consolidation of redundancies due to process improvements.
- **Data**: Project deliverables, messages on handovers, overlap in data/deliverables during the process.
- **Risk and Controls**: Number of identified risks and controls, percentage

of risks mapped to control framework, degree of risk mitigation with identified controls.

As you can see, we are not restricting ourselves to the process view of your architecture, but also touch other architecture views that are relevant in the context of a process. This demonstrates, again, that all views are related to each other and the distinction between process management and enterprise architecture is artificial, and you should avoid falling into the trap of only considering those aspects that are part of "your discipline."

Examples of metrics in the process view

But let's get to the meat of this article – the example metrics of the analysis. The first area is the process view, and the main focus of the analysis is the **effectiveness of processes**. The following aspects should be in your focus:

Every **handover** – either internal to the process between roles, or external process interfaces to other parties (e.g., governance bodies) – creates a delay because the task ends up in a queue for a given role. The ideal state is to minimize the number of handovers, down to a state where one role completes a process from A-Z.

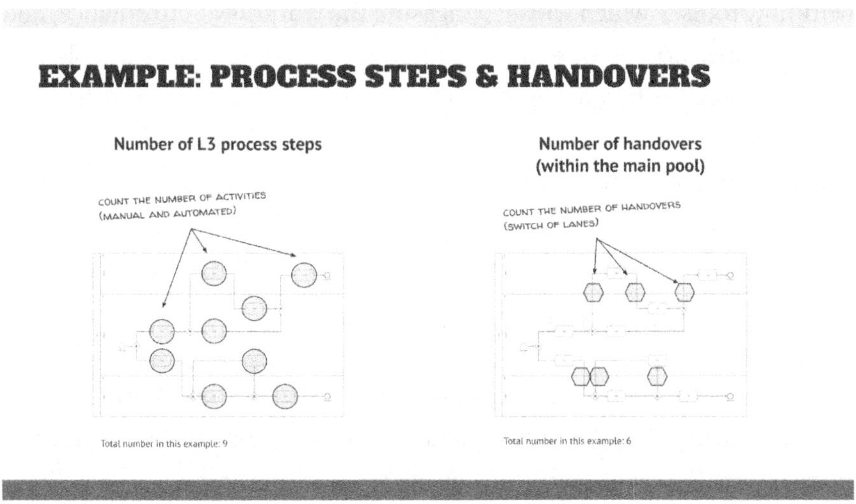

Fig. 30.3: Process steps and handovers as metrics for a process analysis

Switching tools from task to task not only creates an overhead effort in administering and maintaining the tools, but also requires the learning of different tools by the end user (different concepts, etc.). This, and the fact that every tool change causes a "mental switch", delays the processing time of the task. The ideal state would be that a role can complete a process with one tool.

In addition to this, the **manual entry of data** from one tool into another ("swivel chair interface") increases the danger of data inconsistencies, and errors. Ideally, all data exchange will be automated to reduce this danger.

The **number of tools** per process is an indicator of how well the tool rationalization was done in a process area. This can be another indicator for the integration and maturity state of the process.

Multiple official checkpoints and approvals, scattered throughout the whole End-to-End process, create an overhead coordination and consolidation effort that should be minimized.

This is even worse, when the integration/submission is done via unstructured tools, such as email, or unstructured data formats, for example in presentations. In an ideal state, these interactions are supported by workflow tools – which will also measure the workflow performance, and trigger notifications for end users in predefined cases. For example, if a reply is due according to agreed service levels/response times, so that the overall delay in the process will be minimized.

These aspects can be measured, by either collecting the times or counting the number of occurrences.

Example metrics in the application view

The main objective of collecting data and rationalizing applications in the context of the process is the **use of effective tools** and the **degree of their integration**. You might want to look at the following areas:

The number of "real" enterprise tools, that are maintained on a controlled release cycle, versus "End User Computing" (EUC, such as user-created macros, for example), might or might not follow the same principles. They might not be recognized as an official tool with an asset number in the Configuration Management Database (CMDB). This shows an indication of the level of standardization of the tool landscape.

Ideally, all tools are known, registered, and rationalized – which includes a technology capability/functionality analysis to avoid functional redundancies.

Within an analysis, the number of tasks where a user has a "choice of tool" to use, shows the **degree of standardization**. In both cases, the goal is to standardize – and going forward, to automate/integrate – tools as much as possible to improve data quality and minimize risks.

The **number of technical interfaces**, as well as the count of workflows (orchestration of multiple process steps via a workflow engine) show the degree of automation. The idea here is to bring this number up as high as possible, to avoid process inconsistencies (people don't know what to do and find "creative ways" to get things accomplished, which increases risk), and also preserve/create data integrity within the overall tool landscape.

The analysis of the **currently used technology for integration/automation/workflow** might show a preference for a given technology, and is an indicator for a future technology decision (e.g., "buy a new central BPMS, or extend existing workflow capabilities?").

Example metrics in the organization view

Even though organizations and roles are a crucial part of a process, looking at the organizational aspects separately will **provide support metrics** and can improve **efficiencies that go beyond the (limited) scope of the process**. For example, a rationalization and consolidation of roles can lead to reworking an enablement curriculum in the midterm time frame, which will lead to simplifications and savings. Example metrics in the org view are:

- Like in process handovers, the **amount of different org units/roles** within a process increases the potential of delays and waiting time in a process because some information will be stuck in someone's inbox. The ideal state is to track this time and have service levels defined which will be tracked by a workflow engine or process mining automatically (that also sends out notifications if the SLAs are not met).
- Another metric is the **rationalization of org units and roles to find redundancies** which might create confusion ("who is supposed to do what?"). The number of org units and roles – and the reduction of that number — will be an indicator of streamlining. Ideally, there is no redundancy in the responsibilities for certain process steps (e.g., reviews/approvals).

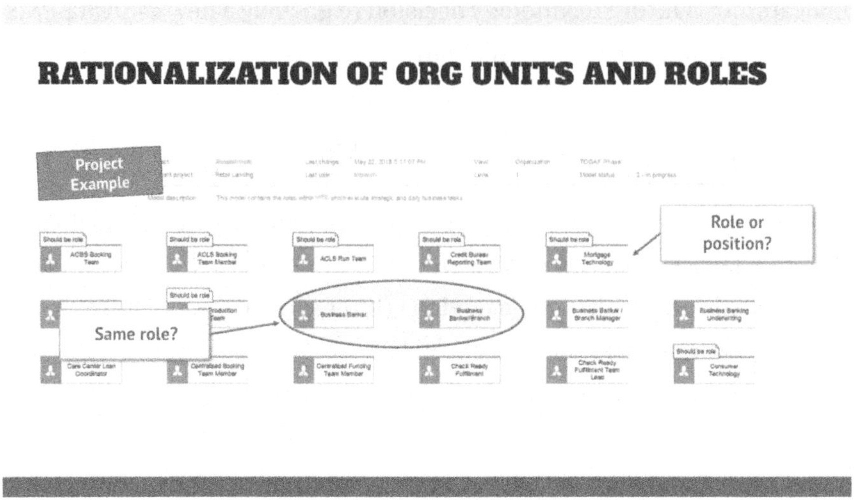

Fig. 30.4: Rationalization of org units, positions, and (process) roles

As a side note, during this exercise the **sequence of role interactions** can be rationalized – an example would be an approval process that requires five different roles to approve sequentially, while that approval could happen in parallel.

Example metrics in the data view

Looking at the invisible data structures and content targets **rationalizing and reducing information requests** within the process, but also **improving the data quality** of the organization as a whole. Example metrics can be:

- Project deliverables are collections of data in a specific format (e.g., a document, model, or presentation) built for a specific purpose. However, more often than not, the same information is requested in a different **format**, and the **rationalization and decomposition of data** is needed to reduce work for re-creating the existing information.
- In an ideal state, all data elements ("entities") of a project are known, and you can track when this information gets created, updated, shared,

or retired. This also enables assigning responsibilities to certain entities and treating them as a "product" going through its lifecycle.

- Typically, data entities also build upon each other, and to improve a process one has to understand **how the information is connected**, so that the upstream/downstream sequence can be created correctly. In an ideal scenario, this data build-up/flow is shown for the whole project, and can be reused/classified for reuse in future projects.
- Understanding the **logical data structure** also will help in **defining the physical data structure** in tool implementations and be the guideline for table creation / field naming while avoiding inconsistencies. The objective here is that changes in the physical data implementation are coordinated and guided by the business use of information (data rationalization).
- Lastly, an **enterprise-wide understanding of the data** that is created/used will reduce the number of redundant, disconnected implementations by providing a baseline of "how things are tied together".

Example metrics in the risk and control view

The last area of the base process analysis looks at (regulatory) compliance. The metrics captured here will give an input on the **degree of compliance** of the process. Example metrics are:

- The **number of risks and the controls** that are associated with processes and process steps will give a metric of "coverage" ("Have we thought of everything? Are all risks mitigated by controls?"). Obviously, you would rather not have non-mitigated risks in your process.
- In addition to this, the **comparison with a GRC framework** (for example, COBIT) will produce information to which degree the risks in the framework occur in the process, what you might have missed, and how well the identified risks cover those.

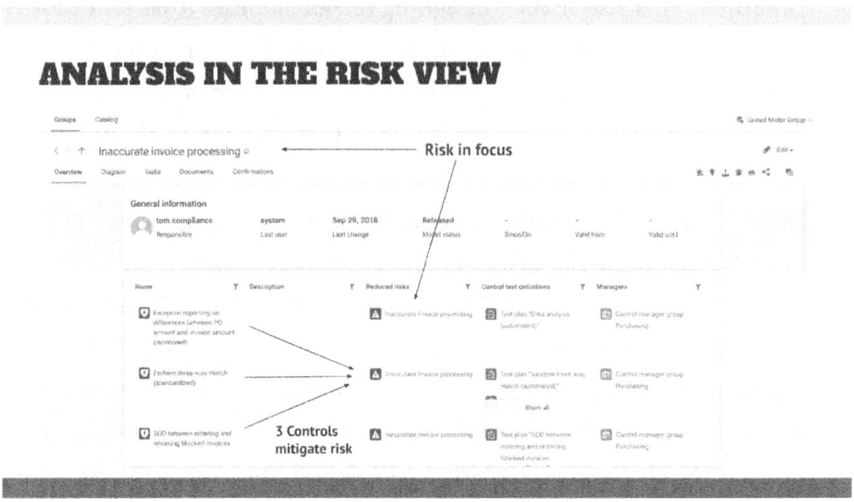

Fig. 30.5: Relationship between controls and risks

- Another metric shows the **relationship between controls and risks**. Do we have a 1:1 relationship or can, for example, a control mitigate multiple risks while still being effective and manageable? Adding a process context can help with rationalizing the risk and control frameworks in the organization.

How do I capture the data, and what do I do with it?

All the different data points shall be collected in a central place, where you can do additional calculations or filter the results by various dimensions, such as org units (to see if the process is a problem just in one area of your organization or a general problem). This will determine your recommendations later.

As a minimum, a spreadsheet can be your initial repository for your data collection – as shown in the project example below. Or maybe your architecture tool provides a means to store this information and display them in a dashboard format. Or you might decide to build a separate database for collecting the data – in the end, what matters is that you have a structure defined that will support a systematic storage of results, and allows you to do further analysis.

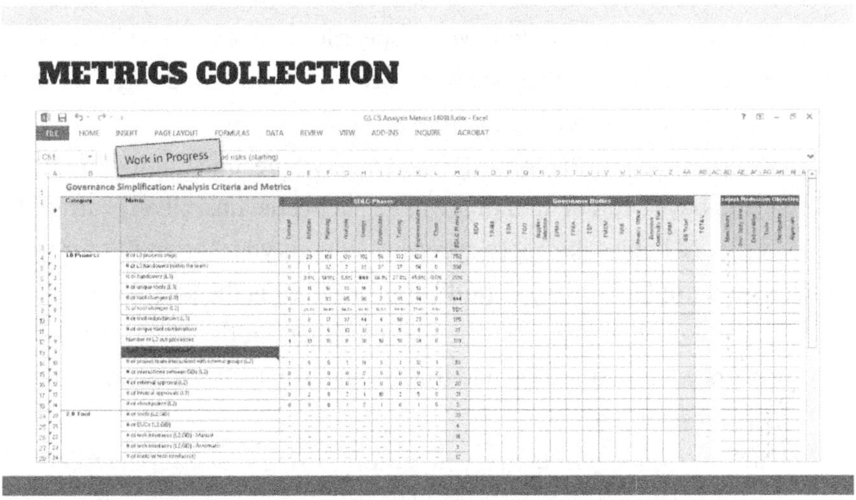

Fig. 30.6: Project example of a "poor man's metric collection"

If in your version 1 of the "base process analysis" package a worksheet is the repository, you should not stop getting this rolled out. Do not try to build "the better mousetrap" (for example a relational database), but save this for a future version of your toolkit.

However, it is not enough to just capture data. You need to create information from it ("what does that mean?") and then generate insights ("what do I do now to improve this situation?"). To create insights from the data in your repository, you want to:

- Search for **patterns**. Take a step back once you have collected the data and check the calculated values. For example, is there a governance unit that always requires a large effort of preparing data and rework compared to others? Or is there a point in the process that stops progress because some contributions by others are consistently not available on-time?

- Develop **hypothesis** and validate them with the data. Once you have a feel for the issues of the process, you might want to develop your hypothesis; for example: "Is the reason for the delay in receiving goods from our vendors a problem with a specific vendor or a specific type of material that is ordered? Are our warehouses built in a way that they don't support the delivery of large goods?"
- Develop and map **KPIs** to the areas. For example, one finding might be that you have a ton of data that will be "delivered on a silver platter" to governance bodies during the process. What is your target KPI for this? Zero and implementing a "pull" approach? Or is the preparation for a major stakeholder group, such as management or the ARB, acceptable? And if so, what defines a group that receives that special treatment and which ones don't?
- Relate this to **pain points** that you capture during your interviews (either while you create the process models, or as part of your analysis). The humans doing these processes for years or even decades know the ins-and-outs like nobody else. But be aware that humans also want to solve problems and might tell you what needs to change ("We just need to replace system XYZ") – this can be true, but I would note the suggestions down and check if they fit into your analysis results later.

As you can see, it is the transformation from **raw data** to **information** into **insights** that is the objective of this phase.

Then you have to communicate your results. When you develop an analysis readout presentation, avoid the trap of getting hung up on the amount of work that you've done, but frame your results in the context of the insights. You might want to use the five areas above as a conceptual framework on a slide if you want to bring an architectural flavor to your presentation and also educate your audience a bit on this topic.

Then develop **action items** for each of these insights that can be measured in the next phase of your project. Your initial readout becomes the baseline for future measurement.

31

Process Simulation

Base Process Analysis is a good start, but it is "stuck" in the past ... the documentation of the as-is, as it was told by a Subject-Matter Expert (SME) or observed by a process-external observer. It does not give you an idea of how you can predict changed processes, behaviors, structures, and so on — enter Process Simulation.

Simulation is typically used in the same phases of the solution lifecycle as the base analysis; either as part of the initial, high-level analysis to figure out if an identified process might have enough potential to be significantly improved, or (more likely) as part of the solution design phase, when the optimal future-state process shall be determined.

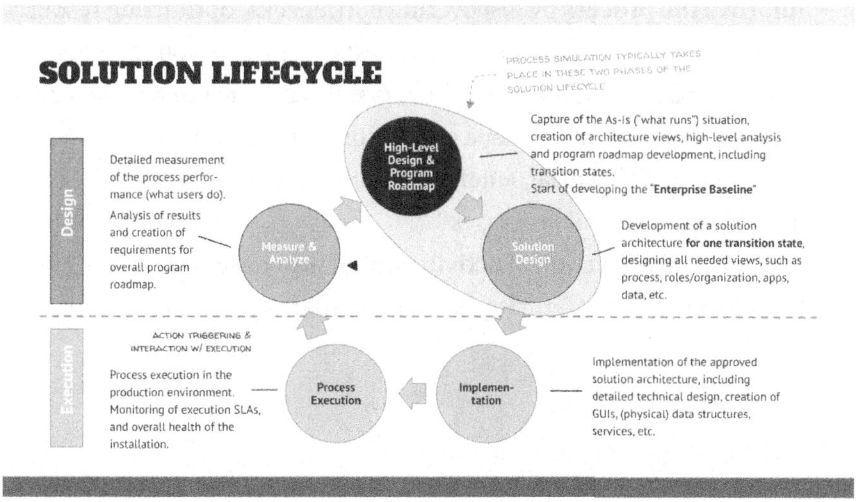

Fig. 31.1: Simulation used in the solution lifecycle

Either way, the objective is to implement a data-driven analysis that will produce a predictable result before any changes to applications, organizations, processes, etc. will be implemented. This reduces risk for the future implementation – spending a lot of money and hoping that the assumptions that were made at the beginning of the process were correct, doesn't seem to be a good and stable approach to me.

In addition to this, process simulation can also be used when creating a business case. The results in resource utilization (people or machines), or the improvement of throughput times/reduction of wait time (one type of "waste" in Lean terms)[72] can be used when creating a business case.

What is Process Simulation, and why do I need it?

Simulation is a first step into a data-driven approach to process improvement. Instead of or in addition to having workshop with SMEs, you can use simulation to determine statistics that will build the foundation of your business case for a future change or implementation.

The typical benefits of using simulation are shown in figure 31.2 below:

- **Improve the process** by discovering bottlenecks, optimizing resource usage, and identifying patterns and best practices in your process
- Identify and analyze potential **process risks** and the **impact** the realization of these risks might have on your organization
- Assessment of potential **benefits** of planned transformations by determining the "optimal processes"
- Analysis and prediction of **activity-based process costs**
- Establishment of enterprise-wide **benchmarks**.

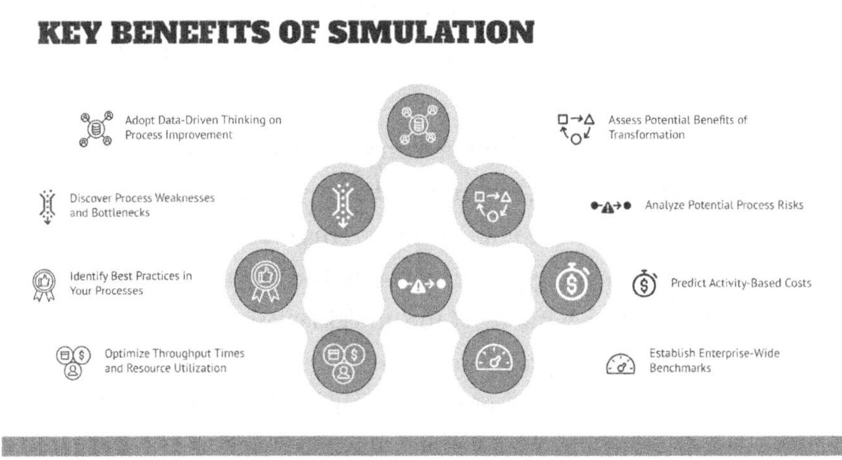

Fig. 31.2: Key benefits of simulation

This also brings up the question of where your Simulation tool should be located in the solution/process lifecycle and the supporting tools – especially since some (pure play) Process Mining vendors include a what/if functionality in their tool sets.

This is a bit misleading on their side because the base piece of every process mining exercise is the underlying data set. It limits the possibilities of a proper simulation to the content of that data set. That means that you can change parameters of what/if scenarios only for parameters of the underlying data set (what is not included cannot be simulated).

This is different from a proper simulation tool that is not confined to the limits of a data set. It allows you to get more / different statistics (such as resource utilization of people and machines), but also allows you to add more detailed information to simulate, such as calendars and schedules. The results of a simulation can be exported to be further analyzed in statistics programs that 6-Sigma folks love to use.

Lastly, a proper Simulation tool allows you to throw out the initial process completely if your analysis points to this result. You can start with simulation, a new process that has been created on a fresh piece of virtual paper.

These reasons and the fact that in a Process Mining tool you just can simulate one process (typically the sizing and cost for a process mining tool will be calculated on the number of processes and cases). While in a Simulation tool that is part of your design tool, you can simulate as many processes as you want — Simulation is just another license/subscription.

What types of simulation exist?

Simulation technology (that is not in your process mining tool) typically falls into two categories that aim for the same result:

- **Discrete event simulation**[73] is an approach to create results by defining parameters for a simulation and then to run these as instances. This means you will define, for example, the frequency of the start event (e.g., 300 times/day) and the simulation engine will run 300 instances of the process for the time frame that is specified (e.g., "for 100 days"). This will then take other parameters into consideration and ensure that all potential walkthroughs are considered. Discrete event simulation is very precise and a "brute force" approach to simulation that works through a queue of instances.
- **Monte Carlo simulation**[74] is a mathematical approach to simulation in

which random samples and probabilities are calculated in an algorithm and deliver a result. Since this is a bit less precise, you will need a large enough calculated sample to approximate the simulation result.

Both methods are not mutually exclusive, and you can find Monte Carlo simulation in aspects of an otherwise discrete event simulation tool for certain aspects (factor variation, such as number of employees in an org unit). This will speed up the simulation for these aspects because the simulation tool does not have to hard run the instances for this aspect.

When you look at the typical simulation areas, you see three major categories:

- **Process-related areas**, including events, processes, cost, and rules in processes.
- **Resource areas**, such as human resources, technical, or capacity resources (machines).
- **Other areas** that cover risks and controls, or the impact of variations in the calendar during the year or simulation period.

Figure 31.3 below shows the three areas and their typical areas of interest in an overview. Each of these bullets can be further detailed. One example is the processing time, when you cannot only define how long a process takes on average, but you can define the minimum and maximum time for each process. The simulation engine will take this into consideration and give you a "best case" and "worst case" result if desired.

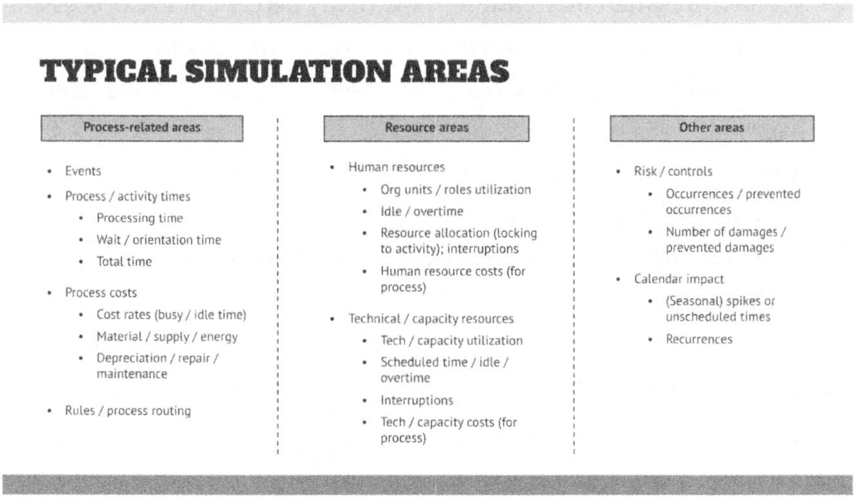

Fig. 31.3: Typical simulation areas

What do I need for a simulation?

Theoretically, you can load any process model that includes events and tasks into a Simulation engine and run it. However, that might lead to a limited analysis result.

Therefore, I recommend to start with four different types of information that will give you a good result that you can define with more details later:

- **Frequency of the process execution**. Does this process happen three times a day, 50 times an hour, or 6000 times a week? This information will be added to the start event, and this determines the number of instances that a discrete event simulation will run.
- **Processing time for each task**. This can be an average time (recommended in the beginning) or minimum/maximum times for the task execution. Once you have gained experience, you might also want to add times for machine reconfiguration or static wait time – literally the time that paint needs to dry in a manufacturing process.

- **Probability when branching** – how many percent of the process instances go left or right when branching in the process flow. This information will be added as a percentage to the connections after the gateway.
- **Number of (human) resources.** This number can be added either for the org unit or the individual roles that are participants in the process, and will help to determine bottlenecks (everyone is busy in another instance) and the utilization of resources.

Figure 31.4 below shows these four bits of information on a process model.

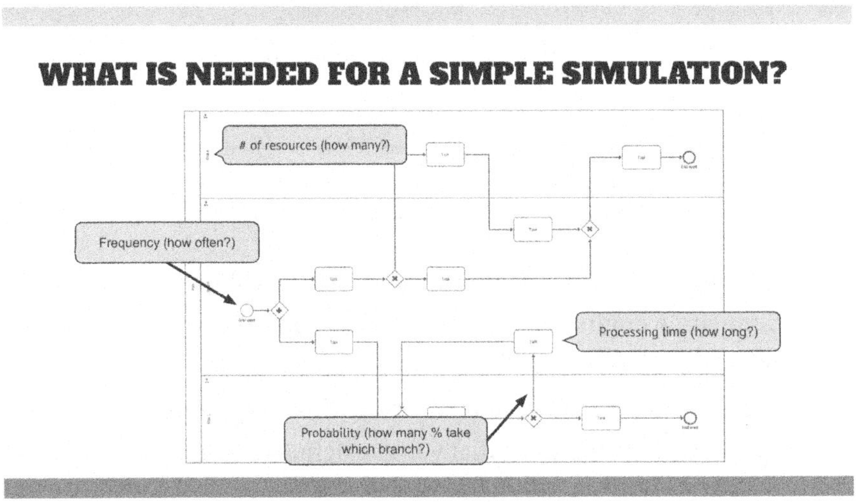

Fig. 31.4: Information requirements for a simple simulation

When running a simulation in a tool, you can see an animation that indicates which step or resource is currently being analyzed. It typically also gives you statistics on how many times this individual object has been triggered.

However, the more interesting statistics will show up when you pause your simulation. Then the statistics will be collected and displayed in a tabular format up to the point where you are in your simulation. You can then select one or more columns in those tables that are of interest and generate charts from the selected data.

An example of this could be the selection of the processing time and the dynamic wait time (time that a process cannot be continued because you run out of resources). The bar chart then will show two sets of bars, and you can easily see not only where a bottleneck is (like the step "Create quotation" in the screenshot below).

EXAMPLE SIMULATION RESULT

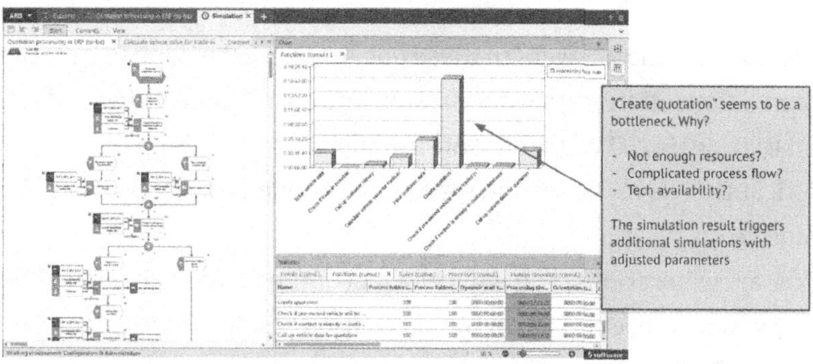

Fig. 31.5: Example simulation result

Now the question is what you will do with this initial simulation result. How do you change your process to get a better outcome? The typical things that I like to look at are:

- Can I resolve the bottleneck by just throwing more bodies at the task (changing the number of resources in the model) and see if it solves the bottleneck or if there is another step now showing up as a new bottleneck?
  This can be used to determine the best number of resources when you

are planning your org changes together with a process improvement project.

- Do I have conditions in the process that will limit resources – for example, if role 1 does step A, then role 1 must do step B. Can I let go of this condition? So that step B can be picked up by another free resource (after the paint has dried, so to speak), so that role 1 can work on a new instance?
- Do I need to change the process, so that my machines (technical resources) can be used with the highest utilization and do not sit idle?
- Or do I need to change the process completely and throw out the old process out of the window and start with a new process?

As you can see, the answers to these questions will be done in iterations – you run a simulation and save the result. You change something and run the simulation again (BTW, turning off the animation speeds up the simulation significantly). Save the second result and compare it with the first. And so on, and so on.

After a few iterations, you will get a feel of the types of changes that you do and which impact it has on the simulation results. Once you are at that point and find a "close enough" result to what you expect as an outcome, it is time to fine tune your simulation by adding more parameters that are mentioned below.

SIMULATION FOR PROCESS IMPROVEMENT

Analysis of as-is processes

- Dynamic correlation of processes
- Semantic correctness of process modeling
- Throughput times of processes with given resources
- Wait (= wasted) times
- Detection of resource bottlenecks and low resource utilization
- Execution of (end-to-end) processes within determined period of time
- Determination of critical paths (time, cost)

Optimization of to-be processes

- Removal of bottlenecks
- Evaluation of process variants for Benchmarking
- Reduction of throughput times and costs
- Increasing number of process deliverables within given time
- Capacity planning
- Optimization of resource utilization

Fig. 31.6: Use of simulation for process improvement

Conclusion

As you can see, Simulation is a good, fast, and cheap way to get an idea of what the "best process" might look like before you start a project to design, implement, test, and roll-out your process improvement ideas.

In a few hours or days, you will have identified the "optimal process" that then can be used as the basis for developing a business case for an implementation project. Just keep in mind that the "best process" can be something different for someone else than you... is it the fastest process execution, the one that uses the least resources, or the one that is the cheapest, for example? Having clarity about your assumptions and targets before you get started with simulation will give you results faster.

Depending on the simulation tool that you use, you can also define "Simulation Experiments," which are pre-configured simulations that you can roll out to regular users who are not supposed to change the conditions of a simulation as described above. This will help you to create another process analysis "product" that you will provide to your organization (similar to the base analysis that we spoke about earlier).

32

Additional Process Analysis Types

We have looked at the whole process of process analysis in this chapter — from setting up the right structure, to importing data, to two forms of analysis that will bring improvement opportunities to the surface.

But process analysis does not happen in a vacuum, just because someone has an idea how to spend their time. It is embedded in the larger quest for insights that lead to an improved execution of the process, which then will also bring business benefits, like an increased capacity of your process ("how many widgets can I create in a given time?"), improved speed ("how do I accelerate the process by removing bottlenecks?"), to increased compliance and other benefits.

But what is the problem?

Let's take a step back and take a look at the issue that exist in how organizations run their processes.

There is a giant lack of **visibility** in all processes that don't create physical goods that you can watch, like a production process, or a retail sales process where you can see the capacity of the process with the naked eye, answering the question of "what is the output of my process?". For most processes, the output is relatively hidden – it is the number of loans that have been underwritten or the number of candidates being hired and ready for onboarding. We also cannot easily see how long a process or its instances takes and if Service-Level Agreements are met.

But these are no new difficulties, and there were attempts to bring in this visibility by using process analysis techniques and BI in the past. Many people have made money by providing services to create visibility and analyzing processes in the past.

The ultimate goal in all of these efforts was (and is) that you change the **execution** of the runtime processes and create results and changes. This can be changes in the process itself, the organization, systems, etc. And you need to be able to measure the performance of these changes as well. Figure 32.1 below shows the problem in an overview.

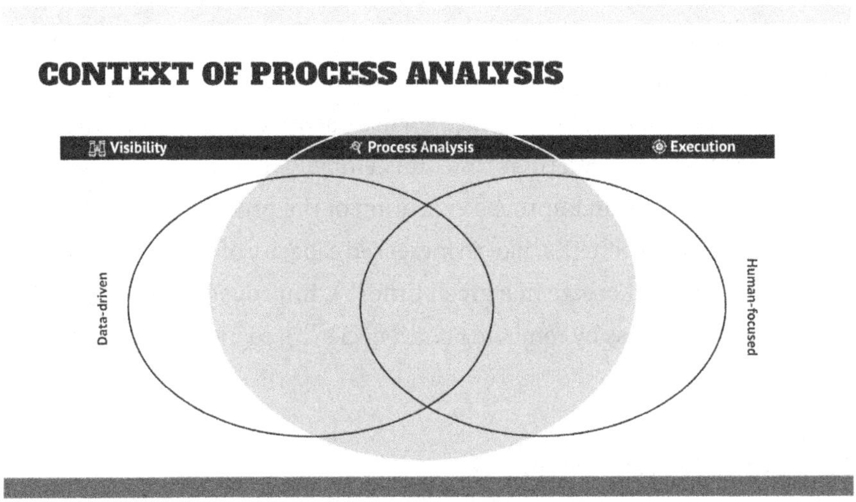

Fig. 32.1: Larger context of process analysis

The key aspect of the difficulty with process visibility is not that we cannot measure process performance — back in the day, people were observing the process with stopwatches and gathered the results on sheets of papers. There is new focus on technology that allows an automated capturing of visibility data that has a much higher quality and objectivity than anything else that has been captured manually before.

Those "mining" technologies allow capturing *all* activities, independent if they leave traces in systems (Process Mining) or are captured by observing unstructured activities on user machines (Task Mining). They also calculate metrics that can give context and insights out-of-the-box, like duration (minimum, maximum, average, median times) and so on, which would take some effort to calculate otherwise.

This allows analysts not only to get to results faster, but they also are based on facts that are in the data set and there is just a limited amount of conversation possible. For example, when the context of a situation is needed when you see an interesting visualization in the process mining charts. This improves the quality of the decisions made significantly because of the subjectivity — "is what I hear from that Subject-Matter Expert the main information, or is he telling me what he wants me to hear?" — is taken out of it.

The big challenge with mining (and other) tools is that they just *support* decision-making, but not make the decision for you. You could trust generative AI tools for this, and we will see a lot of them being built into any tool that we will use soon. But at the core, they are just guessing the next word of an answer by calculating probabilities.

Would you risk the future of your company by trusting a probability calculation that sometimes can lead to "hallucinations" of the AI software? I am skeptical of this and rather see AI support more traditional process analysis techniques as "little helpers" that save time in doing the busywork, but have a human identify recommendations and make decisions on changes.

How do various analysis techniques fit into this continuum?

Process analysis is not a new thing, and some methods have been around for decades and had their heyday in some cases decades ago, but are still relevant. And we will see an integration of AI into the underlying tools, which might make them more powerful and accessible for users.

Figure 32.2 below shows various process analysis methods mapped on the Venn diagram:

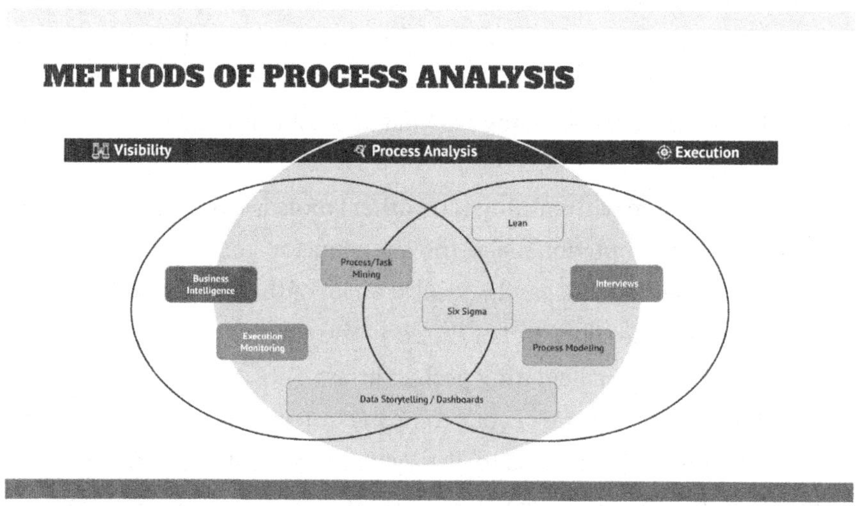

Fig. 32.2: Process analysis methods on the continuum

Let's take a brief look at each method:

- **Business Intelligence**[75] is the practice of creating visibility in how the business is doing by connecting to source systems and visualizing metrics in dashboards. It is the grandfather of what is done now in modern mining tools, but the success of it depends on the creativity and skill of the creator of a dashboard to convey the right information to the

right stakeholder.
One additional drawback of BI is also the fact that mostly lagging indicators are used for the visualization and predictive technology is not widely implemented in the BI tools. They are also not necessarily process-oriented, even though some vendors provide process mining plugins for BI tools like PowerBI or Qlik.

- **Execution Monitoring**[76] is typically built into automation / workflow tools and can be configured to show process metrics. Unfortunately, they typically are focused on the currently running processes in the system (that run in the one system that has this functionality built in, but might have difficulties to include data from other runtime systems in the process) and are not focusing on creating a higher-level summary of all processes and help to identify patterns of process performance.
  Independent of the shortcomings, they provide insights into the execution and have alerting/notification capabilities if something goes wrong. And some runtime tools also allow exporting data that then can be used in traditional BI tools or dashboards.
- **Process/Task Mining**[77] tools fill that gap. They are designed to be process-centric and been fed by multiple systems in the process. However, they might have different levels of analysis tools built in and cannot bring in non-process metrics for visualization purposes.
  Some Process Mining tools can export data to BI tools and have machine-learning/AI capabilities that can give predictions on how cases might end in the future. They can also trigger notifications / alerts and trigger actions in runtime tools. However, they are not necessarily complete regarding their analysis capabilities and Simulation or detailed statistical analysis might be in a different module of a vendor's platform.
- **Six Sigma**[78] is a method to fill this gap. It is meant to calculate deviations in metrics that typically are used to improve quality of processes ("How well is it executed?") and reduce defects.
  Every so often this can be a bit "nerdy" for the regular analyst and go into statistical details that they might see as "overdoing things". Independent of that, the statistical and data-driven approach that Six

Sigma brings to the table will improve the quality of the analysis, and a solid understanding of the statistical approach will be helpful in mining projects as well.

- **Lean**[79] is another methodology that aims to find waste in processes, such as overproduction, defects, inventory, etc. Those shall be optimized and can be helpful to give your mining projects an initial focus if your client has trouble finding hypotheses for (bad) process performance.
  It also comes with various visualization and analysis approaches, like Fishbone Diagrams, that are also helpful in other analysis types. And it is frequently combined with Six Sigma. I highly recommend gaining a good understanding of the Lean method for the average Business Analyst without the need for a formal "belt" certification (Green Belt, Black Belt, and so on).
- For a long time, **Process Modeling**[80] was the analysis technique of choice — visualize the content in a diagram and find consensus how work is done or should be done. Add objectives and measurements to them (current state and future state) and run a couple of reports on the diagrams to find improvement ideas.
  In a more advanced approach to this, tools like **Process Simulation**[81] were used to create a performance baseline and create scenarios to resolve bottlenecks and other unwanted behaviors of the process.
- **Interviews**[82] are the easiest analysis technique – talk with as many people as possible (process performers, process owners, system owners, etc.) and see what "bubbles up" or if your initial understanding of the problems is validated.
  Even though it sounds as if interviews are the least professional analysis technique, they are still needed in many other approaches as well. Interview partners who have done the process for many, many years will give you the context and the details of the process, and thereby deliver the context for your analysis and ultimately the improvement opportunities that you will identify.
  However, the risk with this technique is that you are dealing with people,

and they are highly subjective – they might just tell you what they want you to take away, they might have a biased perspective on the process and its issues, or just think of process exceptions while the majority of process instances might run smoothly within the defined service levels.

- The last technique on the list is relatively new and might be applied to all other methods: **data storytelling and dashboarding**.[83] While dashboards are used in BI and other approaches, they might be considered a technical thing that visualizes information. The risk is that there is no context to the data that is shown.
  The context that you get from interviews will allow you to tell a story with the underlying data and therefore transform it into information and knowledge ... and sometimes wisdom when you can see what you might not want to do ;-)
  Data storytelling is a skill that I have not seen in many Business Analysts or Dashboard Developers, which might approach the creation of them from a more technical view ("Which widget do you want for that data and where on the canvas should it be?").

As you might have guessed by now, to do a proper process analysis, you will have to take pieces of each methodology and combine them to your situation. If you are doing a process mining project, for example, you will have to do interviews with stakeholders. You will also have to extract and transform process data to show them in a dashboard, that is aligned to the hypotheses of process performance that drive your analysis.

What does that mean for me looking at that bigger picture?

Given the information above, it is clear that focusing on just one technique falls a bit short. When you start with your process architecture activities, I recommend to focus on the three steps that we outlined at the beginning of this chapter:

- **Create visibility**: this can be the creation of dashboards (as the BI folks do) or the creation of models and diagrams, which is what traditional Business Analysts have done and then add performance data to these artifacts. In both cases, the objective is clear – get a performance baseline and make it visible because only what you see can be improved.
- **Analyze the process**: combining data-driven and human-focused approaches is the right way going forward. What you want, though, is an analysis result that is aligned to objective goals and not being tinted by the perspective of -important- stakeholders. Your recommendations of opportunities should be based on facts and be reproducible and understandable by another person.
- **Execution**: the best ideas are worth nothing if you don't change things and improve them. This means that your recommendations need to be realistic from an implementation perspective – this includes things like feasibility, cost, or organizational/technical readiness for your ideas. When you implement your change, you should always think of the KPIs and metrics that you have defined and make it an integral part of your solution design. You should capture those and compare them with the baseline metrics that you identified in the previous steps and were used in your process analysis.

33

Real-World Application Summary

As you have seen over the last couple of chapters, creating an architecture view is not (too) hard. You can start small and then build on your first successes.

The key takeaways of this section are:

1. Create different views of an architecture slice — in our example, the functional process decomposition and the End-to-End view. For both, you can find reference content that can accelerate this definition (for example, APQC) and you should strive to keep both views in sync and current, so that other people have an updated information in the (enterprise-wide) architecture tool.
2. Just modeling processes is not enough. There are other aspects, such as applications, org structures, or risk and controls, that might live in other systems of truth and are maintained by people that you might not have worked with. However, you are in the same organization and an organization just has one architecture, so you better work together if you want to do justice to your stakeholders (internal and external, such as regulators). You will see that there will be positive outcomes for both of you.
3. Importing existing content and building technical interfaces to systems of records can be accelerators to your program, and also make sure that

the content stays current, so that others can do their analysis as well.

4. Once you have modeled your processes, there are many forms of process analysis that you can apply. There is not one method that will be the superpowers, but you will have to look at different techniques and approaches to apply to your problem. I showed two analysis types in detail so that you can get started, but there are more to apply.
5. The bigger picture of Business Process Management — and its main issue — is that different people understand different things when using this term. Some think it is just modeling, but I am challenging you to see the bigger picture: you need to get visibility into what your organization does (you can only improve what you can see), and then analyze your processes and identify improvement opportunities (models are a good tool for this, but there are other tools as well). Lastly, the main objective is to improve the process execution — so, don't stop when having identified opportunities: bring them to fruition and measure and analyze the better processes on an ongoing basis.

VII

Before You Go...

34

Before You Go ...

What a ride. I am sure your head is still spinning from all the different aspects of an architecture program that you have read by now.

However, there are a few more things that I like to give you on your merry way to learn more about, where to find information, and overall how to become a part of this exciting community of folks who decided to make their living as architects (or however you name the role that creates all the stuff we spoke about :-)

To help you with these activities, I have set up a website for this book at whatsyourbaseline.com/successful-architecture-implementation. It includes a link to enter the freebies that I mentioned in various chapters of this book. Look out for a link to redeem the freebies and enter the code **SUCCESSFULARCHITECTURE** to get access to them. The QR code at the end of this chapter will bring you to that page.

And what is in for us in the future?

A wise man once said, "Predictions are hard, especially about the future,"[84] so I will not try to look into my crystal ball and tell you what the future holds for us or for the architecture discipline and pretend that I will be right. That might not age well, so take it with a grain of salt when you read my ideas. [85]

But I still like to give you a short outlook from where I stand at the time of writing these words in early 2025:

- Architecture will continue to be a niche discipline (unfortunately) because it is difficult to communicate value if there are no agreed upon measures for success. For example, if you identify savings of $xM because of your analysis, and then you implement a new system that brings this to fruition, the question is: who gets the credit for it? Or is there a credit share, and if so, what percentage goes to which of the two areas? And how do you handle the allocation of credit for benefits in the upcoming years?
  That is why the more IT-focused architecture efforts have a hard time to professionalize (almost nobody outside of IT cares about the cost, and if there is no money for innovation, then you look for "Shadow IT/Shadow AI" solutions elsewhere without thinking about potential compliance issues — just "to get stuff done"). Or even worse, when the fine for regulatory non-compliance is significant smaller than the monetary gains you make by acting in a non-compliant way, then you promote any architecture work as "useless exercise".
  As long as we, as practitioners of the discipline, cannot agree on a shared set of success criteria and promote this, we will not overcome that situation.
- There will be no "AI revolution" in our field (and overall). We are at the top of the first AI tech hype cycle,[86] and people realize that they still have to do work to figure out where to apply AI most effectively ("peak of inflated expectations"). Vendors started to build in AI features at the level of "little helpers", such as importing content from unstructured sources, or summaries/questions of content that is already collected, and I expect more of that.
  There are some ideas of a bigger picture — Software AG is talking about autonomous, democratized, and augmented features and I like these areas a lot — but we are miles away from seeing a lot of that in the tools.
- Speaking of which — there is a market consolidation going on for a few years now. Companies bought Process Mining firms and by now

everyone should have one, and we see the consolidation taking up speed in the IT Architecture sector with Bizzdesign and Orbus making major acquisitions. But there are other "collaborations" going on that might bring together the technical and the business architecture vendors, which in my book would be a good thing. And some big names in the market are up for sale, which one would have not thought two years ago.

But there is a big downside to this: consolidation means fewer players, and given the fact that there are commitments and other constraints in the organizations that heavily influence product management and development (and I have worked for two vendors for many, many years) there will be fewer ideas.

- This, in my opinion, is the perfect opening for an open-source architecture tool that will fill that vacuum (or a "distribution" of open-source tools, similar to what we have with Linux distributions, arguable the most used operating system in the world ... just not on our desktops).[87] And think about it — the current tools were started 20–30 years ago, and I have spoken with numerous people about this, and nobody would build an architecture tool with the same concepts and metaphors today. And on top of it, you can take out privacy and security concerns (open-source software is inherently more secure because more people/everyone can review the code), as well as more cost-effective.

  I can see a new cottage industry growing around this that makes money from services around an open-source software (hosting, training, consulting, etc.) and the barrier for contribution will be lowered. Everyone can contribute to an open-source project, and your specific problem can be addressed faster (sometimes by yourself) than following the "idea management" process of a commercial vendor.
- Lastly, I also have a positive outlook: the benefits of architecture work that you can reap from a structures implementation will still be there in the future, and you can start right now with it. Creating more transparency, identifying gaps and overlaps, designing better future states, communicating more clearly, and all the other things we spoke

about won't go away.
So if you need one advice from me, then it is to not wait for the next tool (the "better mousetrap") but just get started!

But there is more...

Architecture is a discipline that is not done alone. And by that, I mean that you should look out for other people in the discipline outside your organization and connect with them.

Here are a few resources that you might want to check out:

- The **What's Your Baseline? podcast**. Its mission is to demystify Enterprise Architecture and Business Process Management and is hosted by yours truly and J-M Erlendson. We publish on Mondays every other week and have a mix of conversations with guests, about topics of the discipline, as well as things that are relevant if you work in or lead a practice.
  You can find it wherever you get your podcasts (Spotify, Google, Apple, Amazon, YouTube, etc.) or on our website at whatsyourbaseline.com/podcast.
- Also follow **What's Your Baseline on LinkedIn**, which is the main channel to announce new episodes and other things. A lot of the conversations and connections to other practitioners start from here: https://www.linkedin.com/company/whatsyourbaseline.
- A good **aggregator of BPM-specific blogs and podcasts** was set up by Mirko Kloppenburg and can be found at https://bpm.today.
- And lastly, follow or **connect with me on LinkedIn** as well (https://www.linkedin.com/in/rolandwoldt) or **reach out via email** to roland@whatsyourbaseline.com.

I am looking forward to connecting and speaking with you!

whatsyourbaseline.com/successful-architecture-implementation

VIII

Appendix / Examples

35

Appendix / Example

In the previous chapters, you have seen a couple of different modeling ways without me giving a more in-depth look at them.

In this appendix, you find more information about:

- An introduction to the BPMN notation, which has evolved into the standard for lower-level process models, but also as a technical exchange format for automation, process mining, and other applications.
  I referred multiple times in this book to this notation (for example in the process governance chapter), or in the real-life application section. So, here is the primer for you if you haven't heard about it.
- How to model data in an architecture tool above and beyond the limited use case that I described in the Technical Governance phase. Data modeling is a significantly underused view of an architecture so that you can make an impact if you dig deeper into it and apply this methodology to your projects.

36

Introduction Into BPMN

BPMN, which stands for "Business Process Model and Notation," is a standard for creating process models and was created in the first decade of the 21st century by a consortium of industry-leading companies such as SAP, Oracle, or IDS Scheer. The standard is currently managed by the Object Management Group (OMG)[88] and the latest version of the standard (2.0.2) was published in January 2014.[89]

When you follow the discussion in the online about this standard, it seems that BPMN is the cure for all diseases that the other competing standards seem to have, most notably the **Value-Added Chain Diagramming** (with its flavor of Value Stream Mapping), **EPC** (Event-Driven Process Chain),[90] **SIPOC** (Supplier Input Process Output Customer; a popular 6 Sigma model),[91] **IDEF** (Integration DEFinition),[92] or **UML** (Unified Modeling Language, also a standard maintained by OMG)[93] and others. In addition to that – if you believe the hype around the notation that is published by pure-BPMN start-ups, BPM Suite vendors, training providers and others – it also allows you to seamlessly exchange models between different modeling tools and execute (read: orchestrate a workflow) in different BPM Suites without the need to modify anything in a process model or to write code. A beautiful world, isn't it?

Unfortunately, that is not the reality and -IMHO- will not be the reality in the near future.

Where does BPMN come from?

As mentioned above, BPMN is one standard to describe business processes. It was created due to a need in the marketplace to invent a common language to describe the processes and map them to execution logic. With version 2.0 of the standard, it tries to connect two independent development streams to make this happen. Figure 36.1 below shows the development of the XPDL standard[94] -developed by the Workflow Management Coalition (WfMC)[95]- and the development of the BPMN standard, first developed by the Business Process Management Institute (BPMI), and then (after the merger) by OMG.

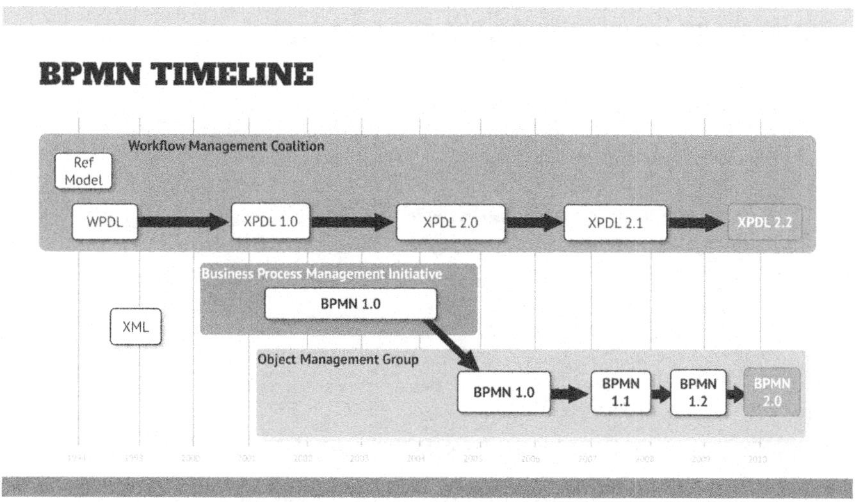

Fig. 36.1: BPMN development timeline until version 2.0 (2011)[96]

The biggest accomplishments in the 2.0 version of the specification is the fact that BPMN has a defined meta model for the first time and an XML mapping that describes how the model information gets "translated" into executable code.

However, this is not reality yet – when you look at the current BPMS implementations there are also competing standards in use, like BPEL

(Business Process Execution Language and its flavor BPEL4PEOPLE).[97] In addition to that, not all BPMS support the full execution of all available objects in the modeling environment. I am sure that there will be some improvement in the future, but as of today, you should have to look first.

BPMN - what's in and out of scope?

The answer to this question can be found in chapter 7 of the spec. There you can read the following:

> **WHAT IS THE PURPOSE OF BPMN?**
>
> Businesspeople are very comfortable with visualizing Business Processes in a flow-chart format. There are thousands of business analysts studying the way companies work and defining Business Processes with simple flow charts.
>
> This creates a technical gap between the format of the initial design of Business Processes and the format of the languages ... that will execute these Business Processes.
>
> This gap needs to be bridged with a formal mechanism that maps the appropriate visualization of the Business Processes (a notation) to the appropriate execution format (a BPM execution language) for these Business Processes.

Fig. 36.2: BPMN purpose definition

This means that **the single purpose is to develop a process model that maps the graphical objects to the execution language**. Nothing more, nothing less. And the standard does this well by providing a myriad of available objects and support for workflow patterns, that are needed in an execution environment.[98]

But what does that mean for the process management effort that you are currently planning or conducting in your company? Is that scope not too narrow? The same chapter also provides the answer to this question and mentions what is out of scope of the specification.

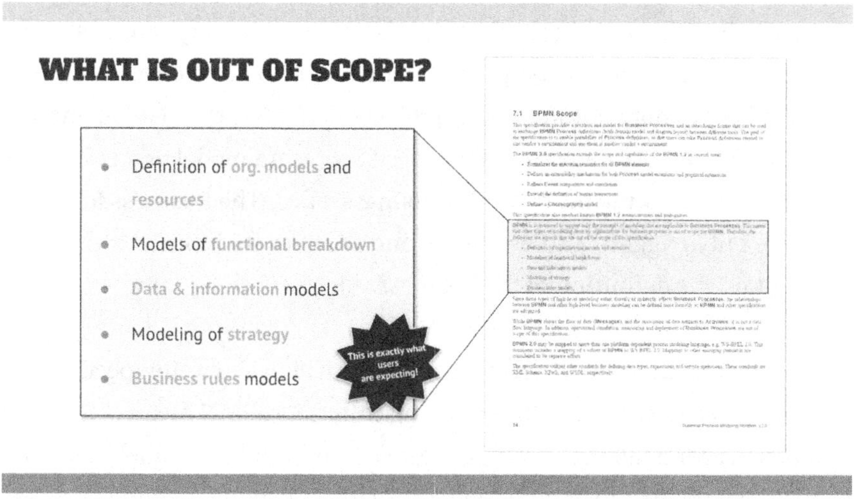

Fig. 36.3 : Use cases not covered by the BPMN 2.0 specification

This means in plain English:

- The specification **does not include objects that represent organizational units, roles, or persons**. Most clients I worked with in BPMN projects use the pool or lane objects for this, but this is a workaround and not the "real thing".
- **BPMN models are meant to be "flat"**, showing one process flow at a time and use subprocesses to "hide" / cluster steps that can or cannot be reused elsewhere, due to the limitations of a sheet of paper. There is no hierarchical breakdown that might be used to create the business architecture layer of an Enterprise Architecture (even though ARIS lets you do this by allowing assignments from VACD functions to BPMN collaboration diagrams).
- Although BPMN has data objects (the Message, Data Object and Data

Store) they are **not meant to be used to create logical data models**. BPMN assumes that all information that is needed in a process is freely available and does not have to be modeled. The data objects should only highlight the most important data in the process and the Data Store, even though sporting the familiar database icon, only represents the permanent storing of data in a system, but is no IT application system.

- BPMN **does not contain models/objects that allow the visualization of a company's strategy** (like a Balanced Scorecard model or an Objective Tree), **nor is it meant to model business rules**. The Business Rule task type only triggers a rules engine and receives its input – it does not contain the definition of the business rule itself.

In my personal experience there is a lot of confusion about the specification and what all the different ways of doing things in BPMN mean ("Hey, did someone order a tall event-subprocess with non-fat soy milk?"), so that these bullet points are undoubtedly the features beyond "simple" process modeling what clients, who believe the BPMN marketing and hype, want to do.

BPMN model types

BPMN is not **one** specific model type -like an Event-driven Process Chain- but is a collection of multiple model types that are connected to each other. This article will show the available model types in the BPMN 2.0 specification and how these are supported in ARIS.

The BPMN 2.0 specification contains some changes compared to the 1.x specification (shown in figure 36.4 below):

- the BPMN process diagram was divided into two model types – the *BPMN Process* and the *BPMN Collaboration* Diagrams
- the new *Choreography* Diagram, and
- the new *Conversation* Diagram.

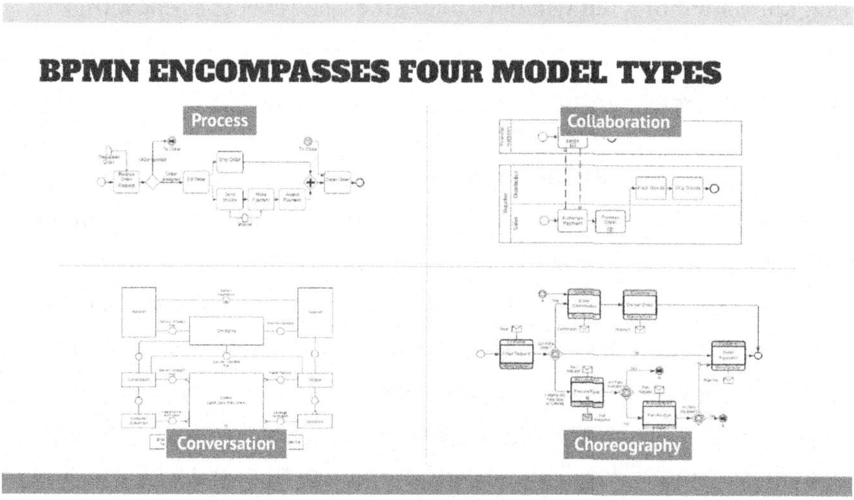

Fig. 36.4: The four model types of the BPMN 2.0 specification

BPMN Process and Collaboration Diagrams

The split of the former BPMN Process into two model types removed the participants and the message exchange and added these object types to the collaboration diagram. The main result of this is that regular BPMN processes *always* happen within a pool, and if you cannot create a subprocess to a process diagram that has pools in it. This is a major clarification that supports consistency in modeling on multiple levels.

BPMN Collaboration Diagrams contain all objects of the Process Diagram and add the message flow between pools (participants). Therefore, they are used to show a high(er) level interaction between participants and the change of control flow between the pools (e.g., one participant is waiting until a response from the other participants arrives via a message flow). Typically, it is best practice to show only one process in one pool in a collaboration diagram and treat the other pool as a black box and assign a corresponding Collaboration Diagram to this pool object.

This also means that you cannot see the participant interaction in Process

Diagrams anymore, but will have to use message events to show this.

BPMN Choreography

A new model type in BPMN 2.0 is the Choreography Diagram. Its purpose is to show the interaction between participants in a different format and concentrate on the message flow instead of the individual detailed tasks of a process. Therefore, new object types were introduced that include sender and receiver within the object (instead of connecting roles to a task or using swimlanes) and connect the send/receive messages to this object.

It also shows the process logic, by using sequence flows and gateways for routing the control flow, but the main purpose of this model type is to show the interaction between the different process steps and participants.

BPMN Conversation

The second new BPMN 2.0 model type – the BPMN Conversation diagram – takes this concept to an even higher level. It shows the conversations between the participants in a 30,000-foot view and shows them as conversation objects (hexagons) without showing the individual message flow, which can go back and forth within a conversation / in one collaboration diagrams. It does not include process logic.

This setup of model types allows showing different aspects of a process for different audiences:

- The 30,000-foot view as an overview by using a Conversation Diagram for senior management or Enterprise Architects, which have to understand the whole interaction.
- On a sublevel, the interaction in a Choreography between process steps for Integration Architects, who are not interested in the minutiae of the individual process step implementation. Or as a Collaboration Diagram, that uses black box pools for process owners.
- On the lowest level, the Process Diagram shows the details of the process

of one participant.

Available BPMN model types in ARIS

ARIS supports four model types, of which three of them are compliant to the 2.0 specification.

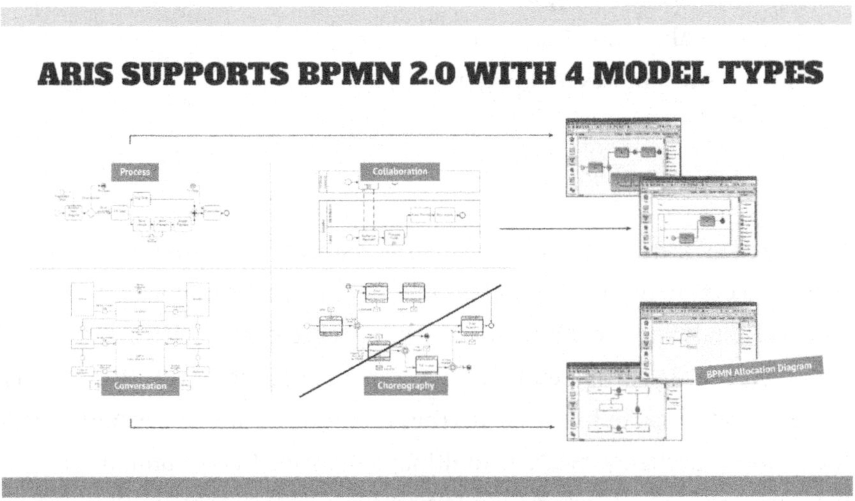

Fig. 36.5: Supported BPMN model types in ARIS

As you can see, only three of the BPMN 2.0 model types are implemented in ARIS, and the Choreography model type does not exist. What was added is a BPMN Allocation Diagram that we discuss below.

Useful additional model types

As we saw above, the BPMN standard does not aim to be a "one-size-fits-all" approach to business process management (even though the BPMN hype creators want to make you believe this), but has a well-defined scope.

Unfortunately, this creates the need for workarounds that will be described more detailed in later articles, but one example are lane objects which have no specific meaning according to the spec. They can describe organizational units, roles, data or systems, for example, which makes it difficult to show / create well-defined relationships in ARIS.

To solve this problem, I recommend using additional model types when modeling BPMN:

- BPMN Allocation Diagram
- Function Allocation Diagram
- "Enterprise" BPMN model types

BPMN Allocation Diagram

The BPMN Allocation Diagram in the current state allows to map participants (pool objects) to organizational units. This covers the majority of organizational needs and is compliant with the specification, which states that a pool represents a partner entity (e.g., company) or partner role (e.g., buyer). However, one popular use is to label the pools with the process name to show the message exchange between multiple processes. In my opinion, this is a fair usage of these objects for non-executable processes, even though you will have to describe this exception in your project conventions.

The current state of the BPMN Allocation Diagram (beta) only allows you to map the partner entity and partner roles, but those objects allow you to assign organizational charts to add more detail. There were more mapping possibilities in the BPMN 1.x implementation, and the "BPMN in ARIS" workgroup is currently thinking about additional mapping functionalities.

Function Allocation Diagrams

Functions Allocation Diagrams (FADs) are a "trusted old friend" from other ARIS use cases, like adding synchronization objects for SAP implementations or adding more details to an EPC to avoid a cluttered layout. The purpose of a FAD is to describe a task (function) in more detail by providing the possibility to connect "real" objects (application system types, risks, or person types/roles to mention a few) to a task. This allows to overcome most of the shortcomings of BPMN diagrams and enables to assign additional model types to these objects. In addition to this, all FAD-related reports that were created can be used in a BPMN scenario.

"Enterprise" BPMN model types

In addition to the "standard" BPMN diagram types, you can also find versions of the BPMN Process and BPMN Collaboration Diagrams with the addition of the word "Enterprise" in the Entire Method. These are enhanced versions where one can, for example, replace the standard lane objects with specific lane objects (roles, organizations, applications, etc.) and reuse existing objects that are used in other views in the database. These are the recommended model types to use, and Software AG just did what other vendors did to take advantage of their database-driven tool setup — a good choice in my opinion.

Objects in BPMN

The BPMN spec contains way over 100 different object types that can be used in the different model types. There are special rules about how to use them, and I will give an overview of the available objects here.

Depending on your use case in perspective, the amount of available objects in the different BPMN models might be a blessing or a curse. Most likely, business people will shy away from the notation and might see it as "too complex", while technical folks who want to implement a workflow system might welcome the possibilities to express even complex situations in a (relatively) simple way.

However, there is a thin line for Enterprise Architects who intend to standardize on BPMN for the process layer to walk on. The openness of the specification in the object definition, as well as the modeling style that allows to express the same thing in multiple ways, make things challenging to define. Since some mappings are not possible or can be only achieved with workarounds.

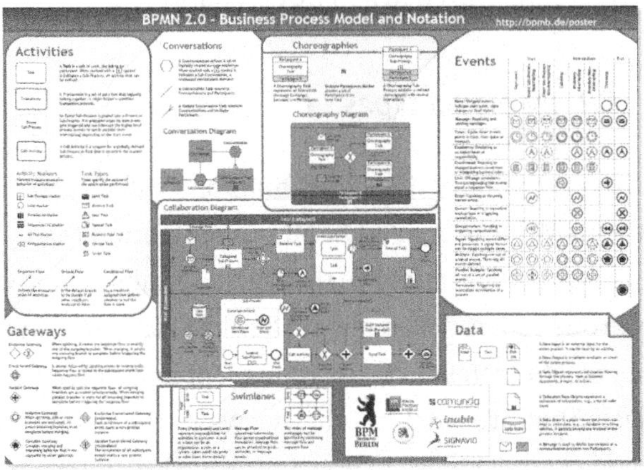

Fig. 36.6: BPMN 2.0 poster created by the BPM Initiative Berlin[99]

While posters like the one above provide a good reference for experienced BPMN users, the usage and combination of objects is not included. In addition to this, the specification clusters the objects in two subclasses to reduce complexity for different modeler groups.

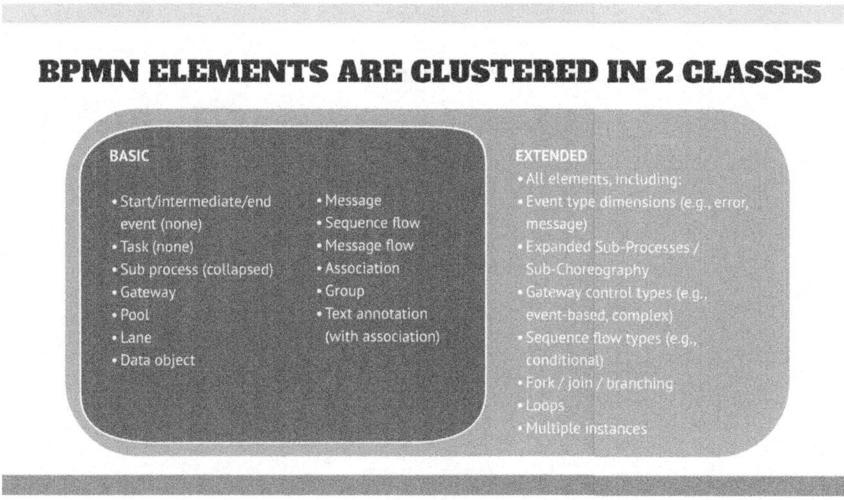

Fig. 36.7: Object subclasses in BPMN 2.0

Objects in the ARIS BPMN model types

The following list shows the available BPMN objects in the current ARIS implementation.[100] Please note that the Choreography diagram is not implemented, and therefore some object types are missing. I have not seen this model type in use, so I assume that the majority of existing users might not miss it at all and use the more commonly used BPMN Process and Collaboration diagrams.

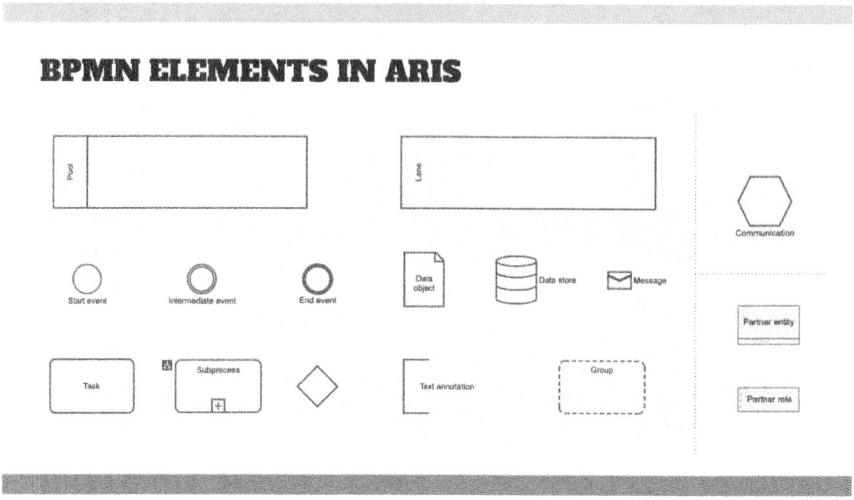

Fig. 36.8: BPMN objects in ARIS

- **Pool**
  (object type: Pool [Participant]; can be used in Collaboration/ Conversation/ BPMN Allocation)
  Pools represent the participants in a collaboration diagram. They are used to show the change of control within the process.
- **Lane**
  (object type: Lane; can be used in Process/ Collaboration/ BPMN Allocation)
  Lanes can represent multiple things, like roles, org units, systems, or data. They are always part of one pool. However, lanes do not have a singular specific meaning, and you have to map them to the "real" objects if you want to do advanced modeling.
- **Event**
  (object type: Event; can be used in Process/ Collaboration)
  An event is something that "happens" during a business process. These events affect the flow of the process and usually have a cause (trigger) or an impact (result). Events are circles with open centers to allow internal

markers to differentiate different triggers or results. There are three types of Events, based on when they impact the flow: start, intermediate, and end.

- **Task**
  (object type: Function; can be used in Process/ Collaboration)
  An activity is a generic term for work that a company performs. An activity can be atomic or non-atomic (compound). The types of activities that are a part of a Process Model are: Process, Subprocess, and Task. Tasks and Subprocesses are rounded rectangles. Processes are contained within a Pool. Tasks can be changed into the different symbols that the spec allows (e.g., call activities or manual tasks)
- **Subprocess**
  (object type: Function; can be used in Process/ Collaboration)
  see Task
- **Gateway**
  (object type: Rule; can be used in Process/ Collaboration)
  A Gateway is used to control the divergence and convergence of Sequence Flow. Thus, it will determine branching, forking, merging, and joining of paths. Internal Markers will indicate the type of behavior control, such as XOR, OR, or AND.
- **Data object**
  (object type: Cluster; can be used in Process/ Collaboration)
  Data Objects provide information about what Activities require to be performed and/or what they produce, Data Objects can represent a singular object or a collection of objects. Data Input and Data Output provide the same information for Processes.
- **Data store**
  (object type: Information Carrier; can be used in Process/ Collaboration)
  See data objects. Data stores save information persistently for this process.
- **Message**
  (object type: Message; can be used in Collaboration)
  Messages show the conversation exchange between pools (= partici-

pants).

- **Group**
  (object type: Structural Element; can be used in Process/ Collaboration)
  Groups are artifacts that can group objects logically.
- **Text annotation**
  (object type: Text Annotation; can be used in Process/ Collaboration)
  Text annotations are a mechanism for a modeler to provide additional information for the reader of a BPMN diagram. But the difference to the standard "stickies" in ARIS is that they are objects, and you can define attributes for them, on which you can report later.
- **Conversation**
  (object type: Conversation [Communication]; can be used in Conversation; Conversation)
  Communications show the interactions of different participants on a 30,000-foot level.
- **Partner Entity**
  (object type: Organizational Unit; can be used in BPMN Allocation)
  Partner entity objects are provided to map pools to org units (that can be reused in other non-BPMN views).
- **Partner Role**
  (object type: Person Type; BPMN Allocation)
  The partner role allows to map roles in BPMN allocation diagrams.

Here are the objects on a nice cheat sheet for BPMN Processes and Collaboration Diagrams:

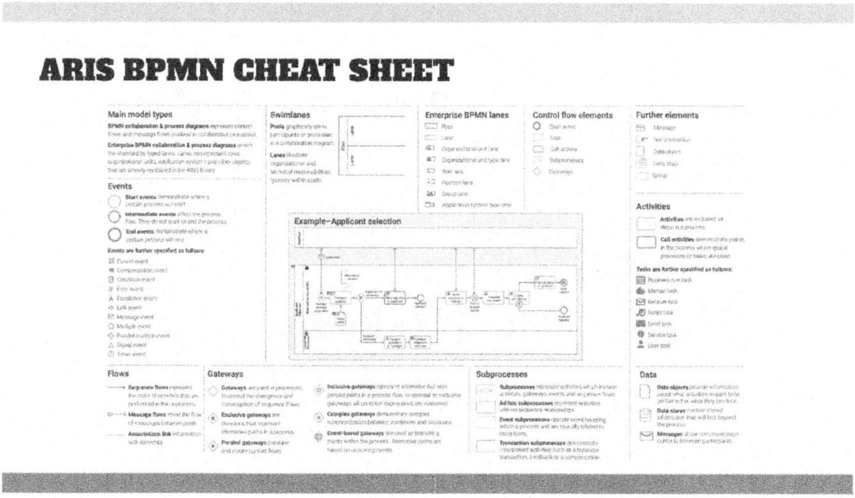

Fig. 36.9: ARIS BPMN cheat sheet[101]

37

Data Modeling

Data models are some of the things that I have seen the least often in projects over the last two decades. Why is that? Is it because it is too geeky, or not interesting enough compared to, say, a process model in BPMN?

This appendix talks about data modeling: the why and conceptual/logical data modeling, the distinction between canonical data model and information assets, the usage of logical data in application models and interfaces, the physical data modeling, and lastly the usage of data on screens and other models.

But let's get started with the immediate question:

Why data modeling?

Data modeling has been long overlooked by business users because it was either considered self-evident (what's a "customer"?) and thereby a waste of time, or was understood as the structure of the physical data fields of databases that stored the data.

But there is a problem with this – without specification of data structure and format, the implementer will make assumptions that might or might not catch what the business person wanted. This is not a good situation.

Data modeling is covering three big questions:

- **What is the data** that we elaborate on – *conceptually*[102] ("what is a customer, and what is it not?") and *logically*[103] ("how should it be implemented without the details of the underlying application known?")?
- **Where is the data** stored, and how is it **transformed**?
- **Where do I ask** for data and **use** it in the process?

The latter is critical today with the heightened awareness of data privacy and the laws that were released, especially in the European Union. By modeling the data as part of your solution design, all ambiguity is taken out of the development process, and the three questions above are easily answered.

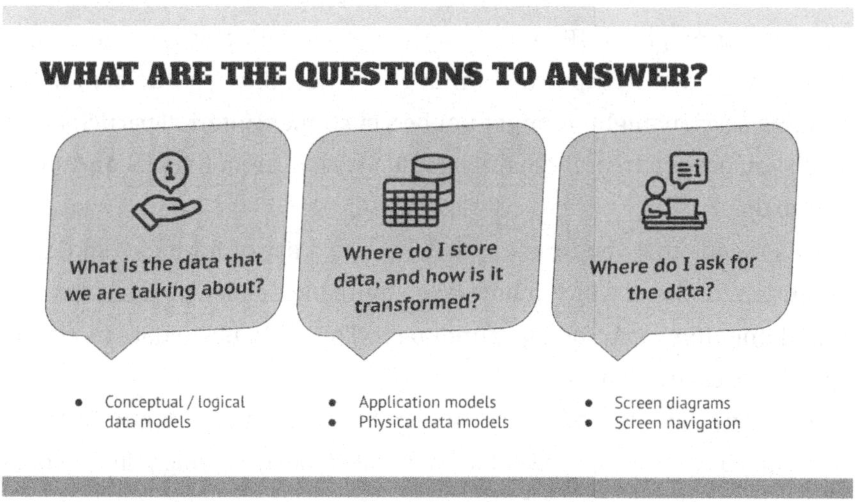

Fig. 37.1: Questions to be answered by data modeling

Conceptual and logical data models – entities and relations

To design a conceptual/logical data model is crucial to get a business understanding of what we are talking about when it comes to the business problem at hand that shall be solved with the support of an application.

Therefore, the definition of "things" is the first step. Remember the "customer" example? I was working on a project that after a couple of weeks came up with this definition: "A customer is an individual or organization that receives a product or service from us, and pays for it".

This sounds pretty straightforward, doesn't it? The benefit of this definition was not so much to have this one term being defined, it was more what it did not define, or excluded. By this definition, an employee who goes through an HR process, such as onboarding, is not a customer, even though the HR department thought so. A training participant in the training session might or might not be a customer (internal participants), so do you have to treat them differently / with different SLAs? And so on, and so on.

This means that the first step is to create a list of "things" and get a definition for them, which follows the "mutually exclusive" rule, meaning that if one thing is A, then it cannot be B. The "things" in data modeling speech are called "entities".

However, there can be a relationship between entities A and B. These relationships can be mapped out in an IE Data Model, and might look similar to the upper half of figure 37.2 below.

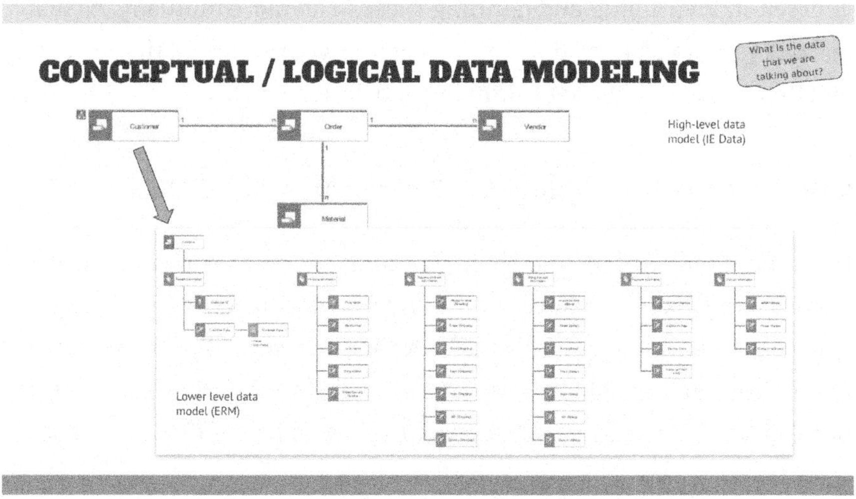

Fig. 37.2: Conceptual/logical data modeling

As you can see, there are relationships between "Customer" and "Vendor" via the "Order". Without the order, there would be no relationship in this scenario. However, there could be multiple indirect relationships between Customer and Vendor, for example, if an invoice has to be paid.

Once you have brought the entities into your architecture tool, and modeled the relationships, you can further specify the types of relationships between the entities. Is it a One-to-One relationship, which means a Customer can have only one Order, or is it a One-to-Many relationship (which is more realistic because a Customer can have multiple Orders, even from the same Vendor)? This information can be added to the connection, and either displayed as characters, as shown above, or with symbols on the connections. This is called "cardinality" in data modeling language.

Overall, it does not make a big difference, in my opinion, if you use characters or symbols. If you model it like above (using characters) you might make things easier to understand for other business users, who might be too shy to ask what this weird chicken scratch on the connection means.

Furthermore, keep in mind that you do not think about the specification of the entities at this point, and also do not try to categorize entities or limit yourself in identifying and defining these. The more mutually exclusive entities you can identify that are relevant for your solution, the better it is and the more complete your thinking and design will be.

Conceptual and logical data models – attributes

Once the main entities and their relations are defined from a business perspective, it is time to dive deeper and describe the next level of detail – what are the entities comprised of? For this, you can create an assignment to an Entity Relationship Model (ERM), and start describing the entities from an attribute perspective (shown in the lower half of figure 37.2 above).

- While doing this, keep in mind that we are still looking at the **business perspective**. Do not try to find the correct field names and add them here – most likely you will discover that the piece of information that you are describing, such as the name, is also used in other systems and the field name over there might be entirely different.
- You can group the attributes logically using **attribute type group objects**. These are of no interest for developers, but will help you when reviewing your data models with business stakeholders.
- When **defining the attributes** you might want to define key, foreign key, and the standard descriptive attributes (note the different symbols in the light blue objects above). But do this from a business generic perspective (again), and do not focus on a specific implementation (yet).
- After that you can start **defining the field formats** – text, integer, date, and so on – and in the case of enumerations (pre-defined values), you can add the enumeration as an object (the gray symbol in the first "tree branch"), and add the values in an attribute that can be shown in the model if wanted.

Doing the latter two additions to an ERM diagram changes the data model from a conceptual to a logical model according to some methodologists. In reality, it is just another enhancement of the model to another level of detail, which will help later in mapping exercises, as well as the implementation in bits and bytes.

Canonical data model vs. information asset view

When defining your meta model (as I have shown for an architecture tool implementation where we looked at the process view as an example), you define levels and views. One of these views is the data view, and it typically includes three aspects:

- The conceptual / logical decomposition (the "canonical data model") that we took a look at above.
- The use of a subset of the canonical data model in the information asset view.
- The physical data models, describing the implementation of the data model in tables and fields.

The greatest challenge that I see in projects when data shall be added to other models is the use of entities (the "things" that are defined from a business perspective) vs. the use of clusters (aka "entities in context"). Depending on your tool and/or chosen notation, you only might be able to use cluster objects -for example when using BPMN models in ARIS (the 'document' symbols specified in the notation).

This is fine, but it also forces you to add another modeling, as shown in figure 37.3 below:

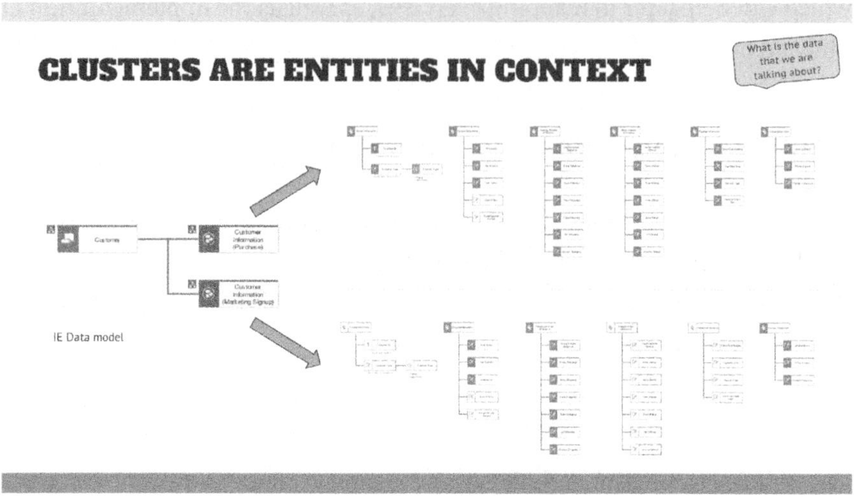

Fig. 37.3: Clusters in the (logical) information asset view

Here you see an entity "customer" that we've described in the previous article: personal information (name, day of birth, etc.), shipping and billing addresses (city, zip, street, etc.), or contact information (phone number, email, and so on).

What you also see is that this entity is mapped to two clusters – one that describes the customer information in the context of an order process, and one that shows the customer information in the context of a marketing sign up process. In the examples to the right, I grayed out the attributes that are not used (in a real-life project I would have deleted the grayed out objects FWIW), and while both clusters share some attributes – like parts of the personal information, shipping (home) address, and contact information – there are specifics to an order that are not used elsewhere (billing address information for example).

This means that you can be very specific in the information definition / needs in a specific context, while on the other hand, you can build your canonical data model by layering the cluster models on top of each other, and aggregate the attributes in the assigned ERM diagram for the entity.

This will take out the big effort upfront "to get things right" in the first place, which might be a reason why so few projects do data modeling.

Applications and data (from a logical data model perspective)

Another way to use clusters, besides using them in processes, is to add them to application collaboration diagrams, which show the logical interfaces between two application systems (types) and which data will be sent to this interface.

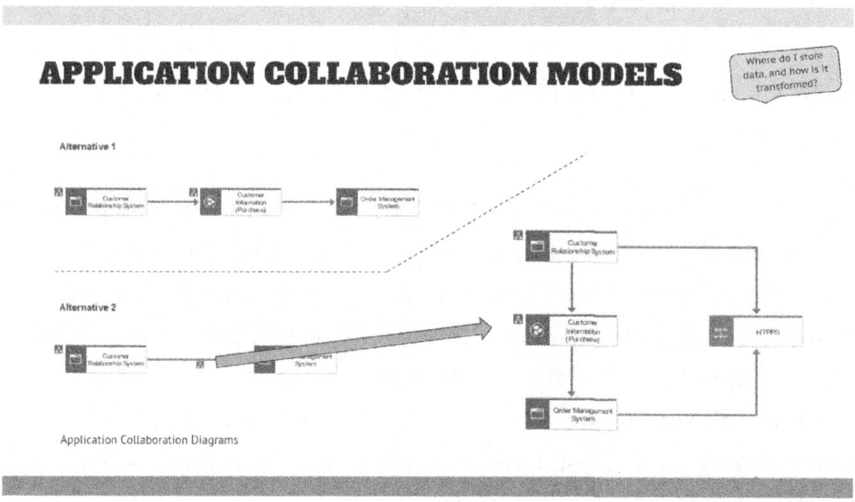

Fig. 37.4: Applications and data in application collaboration models

Depending on the scope of your project, you might do this in two ways:

- model it like alternative 1 and put the cluster between two applications, or
- connect the two applications and add an assignment to the connection.

Alternative 2 has a couple of advantages – firstly, it declutters the application collaboration diagram, which helps to read large application landscapes significantly. Secondly, you can add additional information to the interface description, as indicated by adding the protocol object to the assigned model.

And lastly, you might run into the scenario that a large application with multiple modules might have interfaces with another large application with multiple models. Since you can create multiple assignments to the connection, you can specify in the assigned model name which scenario you are describing when modeling the interfaces.

As with the above example, this also makes your model more resilient to future changes because you will just have to change a lower-level model instead of changing the application landscape every time.

Use of logical data model objects when defining interfaces

As we've seen when we spoke about data model creation as step 1 of the technical configuration, you can use logical data attributes also to map to each other for data interfaces that will detail the modeling from above. This will make it clear how attributes are mapped, even though there might be changes in names in the other application.

You can also create an assignment to a matrix model to the cluster objects shown above so that you have this specification always available when looking at the cluster in other contexts.

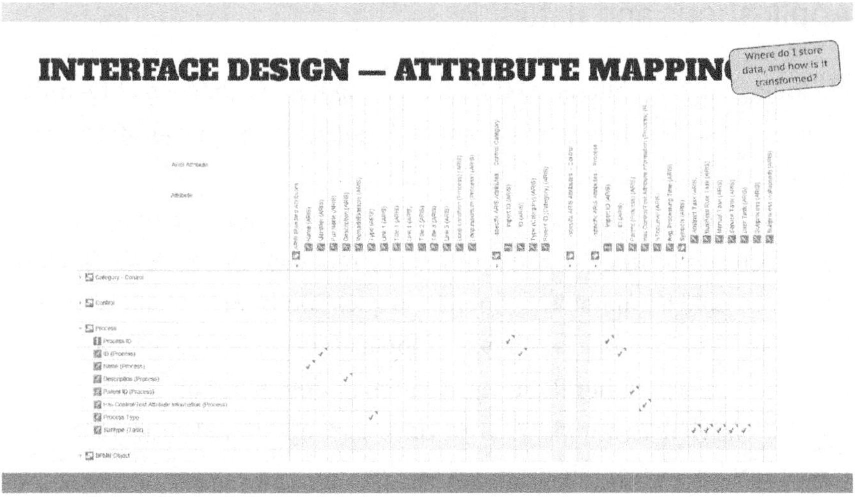

Fig. 37.5: Attribute mapping in interface design

I have seen matrices like this using physical field objects for the description of an interface – and it might be attractive to do this – but I would recommend doing this design using the logical attributes. That is because the technical implementation might change in the future (for example, when migrating an application to a new database system), and you have no insight into this, and the implementation project might not be aware of this artifact, which renders it useless.

The result of this will be that any analysis that you run on your database might now give you a "wrong" result, which is what you want to avoid.

This situation can be overcome when modeling physical and logical mappings, as it will be shown in the next article. The matrix above will stay correct and relevant as the business requirements for the interface, and changes to fields or table structures will be shown elsewhere.

Applications and data

We looked at how data is used in describing data exchanges at technical interfaces and how it might be mapped / transformed. The question now is how data will be stored in the applications.

The first thing to answer this question is to determine if your application has a built-in storage (database) or if it relies on an external database system to hold the data. This opens the question if we are talking about a logical application model or a physical application model (using "application instance" objects rather than "application system type" objects).

Either way, it is a good idea to have two additional models in your repository for each application system type:

- A profile of the application: which technology does it run on, who is responsible, who is using it, what additional documentation is available, and
- The mapping of the logical application (type) to its instances, which then can be used in physical diagrams, such as network diagrams.

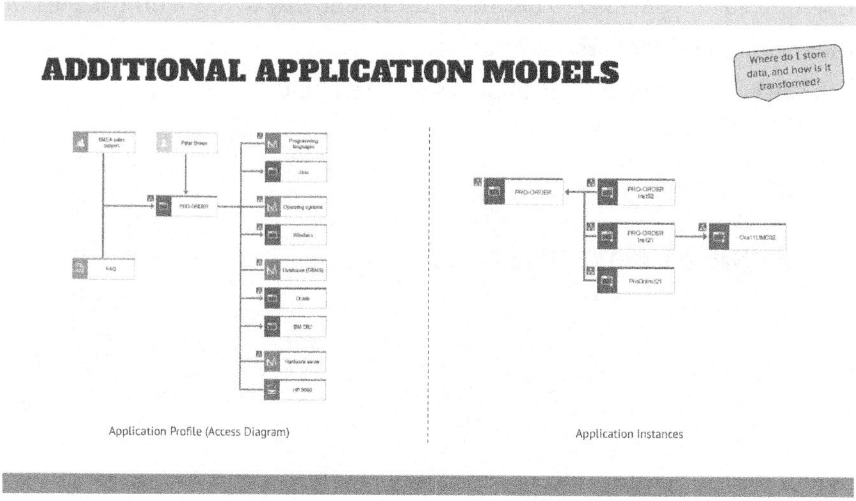

Fig. 37.6: Applications described in additional models

On the right-hand side in figure 37.6, you can see that the PRO-ORDER application has three instances and that instance 21 is using an Oracle instance "Ora11UMC02". This is consistent with the logical profile on the left that shows that the PRO-ORDER application can run on either an Oracle or DB2 database systems.

One thing you can also do is to create an application profile for each instance, depending on what your stakeholder information needs might be.

Physical data modeling

At this point, we get into the physical data modeling, and the way to do this is to create two levels of physical data models, as shown below.

- If the application has an internal database, then you might want to create an assignment to a table diagram directly to the application system (type) object.
- If the application uses an external database system (as shown to the right in figure 37.6 above), you might want to go to the instance modeling.

In either case, you need to define a consistent way of modeling as part of the meta model conversations so that your artifacts are comparable, and you can create standardized analysis and dashboards.

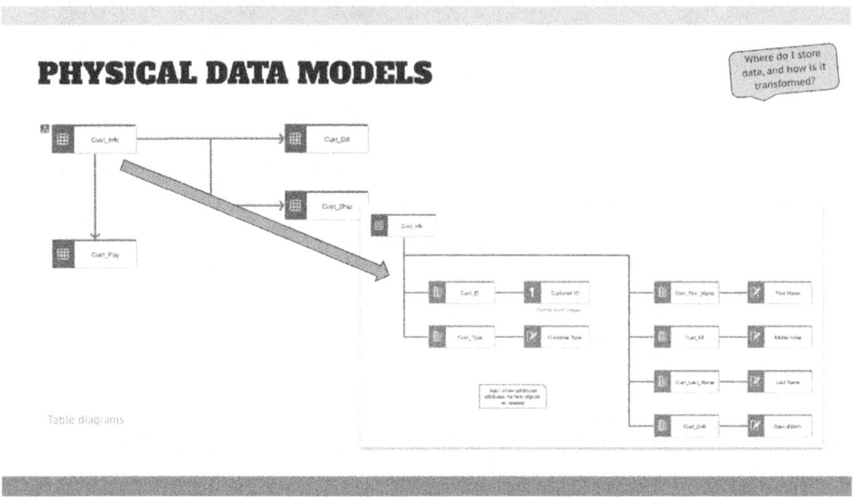

Fig. 37.7: Table diagrams describing physical data (and mapping it to logical attributes)

The first level is a table diagram that shows the tables of the application and how they relate to each other. The example above is a very simplified version, and in projects I have seen far more complex diagrams. What you will see in those pretty quickly are tables that are essential and frequently used.

Once you have created the high-level table diagram, you can assign a level two table diagram to each table object.

In this, you then can model the fields, and other objects, and then map them to the logical attributes. This does multiple things for you:

- You have a **completeness check** – all your logical attributes should find a home. If not, you either have a gap in your physical data design (missing fields or different relationships might be needed if a crucial logical attribute is stored in another table of your application, or even a different application that an interface needs to be built to), or you might identify an inefficiency in your database table design when you see unused fields.

- Another benefit is that you will see that some **fields got reassigned over time**, and their name does not reflect the content that shall be stored there – represented by the logical attribute. This might trigger an **analysis** of the data stored in various applications are really consistently handled.
- As part of the field attributes, you can define the **field format** (e.g., "Text with x characters length") and then **compare this with the business requirement** for formats that were identified when modeling the logical attributes.
- Once you have modeled these relationships for multiple apps/tables you can **aggregate** them and define your business requirement more clearly going forward because you know that you have limitations in your applications, and the lowest denominator might define the maximum amount of available characters.
- You created a **traceability** that is valuable in regulatory obligations scenarios, such as data privacy regulations.

As with the applications above, you can create additional models when using table diagrams – you can create views of related data ("by topic") to make it easier to follow for stakeholders or in workshops, as well as distinguish an "abstract" modeling ("by type") or going into the specific field ("by instance", for example, when you say "we typically have the fields specified like this -format-, but in the database on instance x we extended this format to 2 more characters for these and these reasons").

Data lives in the context of processes

Besides using data in application models as shown in the last article, typically you see data being a part of processes. It can be shown as inputs and outputs mapped to process steps. In BPMN for example these will be shown as "document" symbols, but typically these are clusters and can be treated as described above.

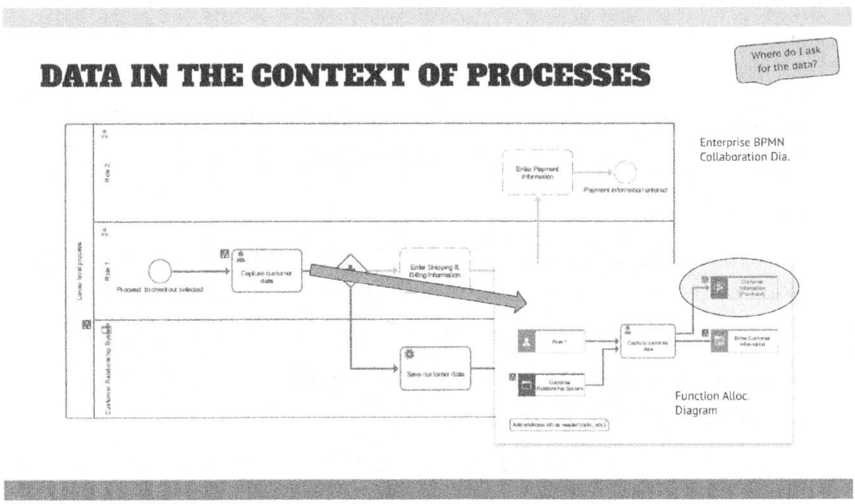

Fig 37.8: (Logical) Data used in process models

Another way to model data is to add the clusters directly to an assigned model to the process step (in a Function Allocation Diagram). This provides the benefit that the main focus of the process — showing the control flow — is not cluttered by other objects.

Modeling it this way, which by some tools is implemented by default when creating relationships using the properties dialog, has the benefit that you can be more specific in your modeling by using object types that are not included in the process notation (for BPMN that would be applications, risks, controls, etc. for example). You can also use different connections to model

RACI connections of roles to tasks, while avoiding a complicated modeling of the process because a swimlane only has the semantics of "carries out" but does not allow modeling a second connection, such as "contributes to".

Data in screen diagrams

As you see in figure 37.8 above, there is also a screen object (the green object to the right of the task) modeled. Typically, they are used in SAP implementations, where screen objects represent the ERP transactions.

But this is just one way to use screens – the better way to use them is to assign a screen design model to them and do a mock-up with data mappings to them.

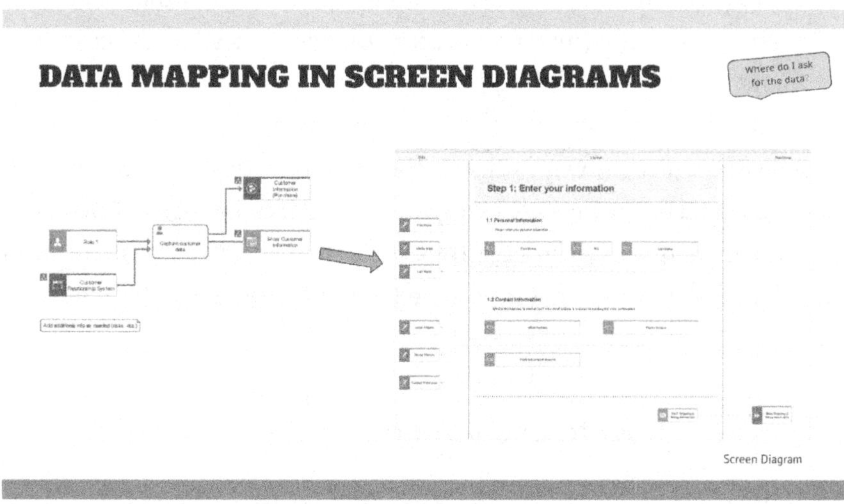

Fig. 37.9: (Logical) Data in screen diagrams

In the example above, you see a three-column layout: a data column on the left, the screen elements in the middle, and a function column to the right.

The main purpose of this diagram type is to do a mock-up/low-fidelity design of the future application screen. Therefore, a good architecture tool gives you objects that represent text fields, dropdowns, radio buttons, checkboxes, and so on, replacing the need for a wireframe tool like Balsamiq.[104] It also allows you to structure the sections of the screen, as shown with "personal information" and "contact information" in the graphic.

On the bottom of the mock-up, you see a button object that could read "next" or "enter billing information".

The next step now is to two things:

- On the left side, you see the attributes from your logical data models being reused and mapped to the screen objects. This answers the question "where do I ask for this data?".
- On the right side, you see a function from the underlying process being mapped to the button, indicating the next step. This brings together the outside-in perspective of customer journeys and the inside-out perspective of the (internal) process that delivers what the customer wants.

But screens are just one way to answer the "where do I ask for data", another example would be forms where you can do similar mappings in Document Structure models. However, I typically prefer to model the data in forms in a regular ERM diagram that is assigned to the cluster object (as shown in above).

Excursion - how to use screen objects elsewhere?

Using the screen objects as shown above is one way to do this. However, screens are one of the most underused objects besides these two use cases. It can be pretty powerful to use them to design the user flow in your new (web) application development, and is a perfect companion to the customer journey maps that you might have created elsewhere in your architecture tool to capture the "outside-in" perspective.

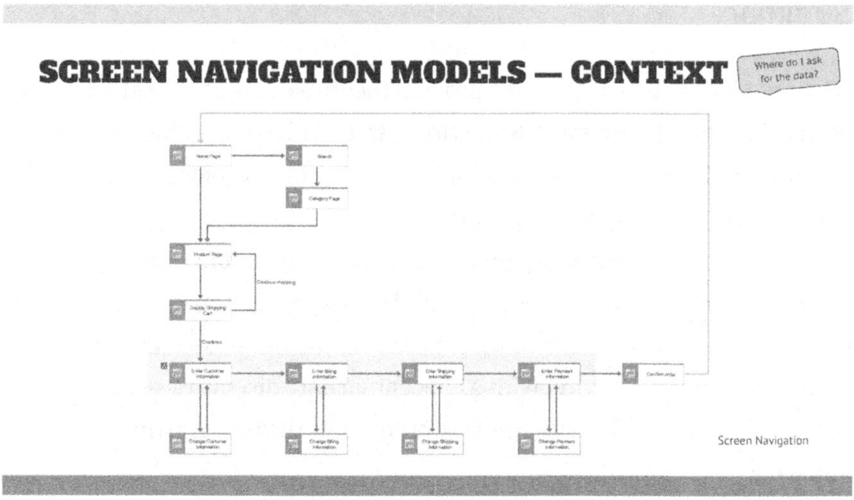

Fig 37.10: Screen navigation models — context and flow of screen diagrams

The mock-up example shows a typical screen flow in an online shop – a user goes to the homepage and either finds what she is looking for directly as a suggestion, or will use the search functionality to get to a product page via a category page.

Once the product is added to the cart, the customer data example that we've used through this series kicks in and the steps for adding shipping info, billing info, payment info, and contact info are shown in the gray bar above.

One thing while creating this customer happy path was to notice that the information might have to be changed, therefore four "change" screens were added. In addition to this, the question of where to go after a product was bought came up, and screen navigation models are a good tool to have a discussion about where you want to steer the customer after an action has been accomplished.

This, in addition to the detailed screen design, will then also help to identify where to make changes in a future release when new information requirements need to be added to the application.

Summary

When modeling data as part of a project, the three questions that we looked at in this chapter come up all the time, and it might be enhanced by data privacy requirements. I have also seen that plenty of projects struggle with this, and don't know where to start.

Unfortunately, there is no reference content available that would have proper definitions that an industry would have agreed upon to my knowledge (please correct me if I am wrong). Therefore, the three-step approach that we discussed here, starting with a conceptual data discovery session — and don't look at tools for that in the beginning — is the best starting point that you have.

Credits

Every book is an assembly of content from "somewhere". Here are the credits for the external material that I have used in this book.

- Cover — the main photo of the title is by Kelly Sikkema on Unsplash,[105] the collage is done by me
- All tool screenshots are taken using the ARIS platform by Software AG[106]
- Icons in graphics are from The Noun Project[107]
- Additional input used in graphics in this book are linked in the corresponding end notes

Praise For "Successful Architecture Tool Implementation"

- "This book has great value. I know I would have loved a book like this when I started my career."
  Marc Mendelson, Business Strategy, Transformation & Customer Experience Executive; Big 4 Consulting Director[108]

- "This book offers a valuable in-depth exploration of Enterprise Architecture, addressing the Why, How, and Who."
  Hanneke Loefs-Mos, owner OLCAMA Consultancy, EA/BPM Consultant[109]

- "'Successful Architecture Implementation' is skillfully written, combining knowledge and experience with a touch of humor, which takes this book from what could be a boring textbook to instead being an engaging read! Roland effectively highlights critical aspects of the implementation journey, ensuring adoption and traction, along with practical 'how-to' advice and some humorous lessons learned. Readers will find many valuable insights, regardless of their stage in the architecture implementation journey!"
  Breanne Casteel, Enterprise Business Architect[110]

- "Roland's book is easy to read and understand. It put me directly into implementation mode and gave my perfect arguments for my top management to convince them of the need to implement all the EA and BPM ideas I am fighting for years."

Martin Holling, Head of Process Management[111]

- "Roland operationalizes the art of enterprise architecture and presents a hands-on guide to mastery of the domain. Anyone starting or running an architecture program in a modern business will gain valuable insights into proven methodology. He covers the entire architecture process from inception to establishing the architecture capability in the organization and draws from a wealth of experience. His hands-on tips for the effective employment of architecture tools are just the cherry on the cake."
  Marcus Zschuckelt, Consultant for Enterprise Architecture and Business Process Management[112]

- "Most organizations underestimate the critical role architecture plays in their performance, and even those that recognize its value often struggle with execution. Roland's book is an essential guide for anyone looking to transform architecture from theory into a powerful driver of change."
  Michael Schank, Author of "Digital Transformation Success"[113]

- "I've read so many books offering business practice advice, but they all lack something that 'Successful Architecture Implementation' has in spades: a specific path for success. Instead of waxing poetic, Roland takes a pragmatic approach, clearly laying out the case for an architecture as a strategic capability, the ingredients of a good practice, the approaches to maximize your chances of success, and tangible examples to follow.
  This book lowers the barrier to entry for passionate advocates, equipping them with key tools and techniques to explain the practice and get stuff done. Highly recommended!"
  J-M Erlendson, Global BPM Evangelist

Notes

PREFACE

1 Take a look at the clients I worked with and the topics of the projects on my website at https://woldt.de.

2 ARIS stands for *AR*chitecture of integrated *I*nformation *S*ystems and was first developed as a method at a university in Germany in the 1980s and 90s before a tool was built. You can find more information about the tool at https://aris.com.

ACKNOWLEDGMENTS

3 I also highly recommend a premium subscription to the "LanguageTool" (languagetool.org). A lifesaver for every author — and I always get a kick out of seeing the "Picky Mode" switch in this German software. This is an oxymoron!

4 https://www.whatsyourbaseline.com/podcast/

THE CHANGING ROLE OF ARCHITECTS

5 In case you haven't read it in the preface — in my humble opinion, "architecture" is the true architecture of an organization. We see many attempts to divide this discipline into different "things", such as "Business Process Management", "Business Architecture", "IT Architecture", or (shudder) "Enterprise Architecture" — when IT Architecture Management is meant. This is the wrong way forward: an enterprise/organization has only one architecture at a time, and that might be "organically grown" or implemented following a plan. For obvious reasons, I think that the second is the better way of doing things.

6 I have heard that term for ARBs many times from the projects who "didn't make it" through their process. However, after the end of this book, you should have a "Committee of Know" as ARB :-)

7 For more about that topic, see here: https://www.whatsyourbaseline.com/2021/12/why-consolidate-your-management-of-change-processes/.

8 https://www.scaledagileframework.com/

9 https://www.house.gov/the-house-explained/branches-of-government

10 https://www.whatsyourbaseline.com/2021/12/episode-11-governance-design/

WHICH CAPABILITIES DO I NEED TO BUILD MY PRACTICE?

11 For more, see my article for cio.com here: https://www.whatsyourbaseline.com/2017/10/enterprise-architecture-the-scarce-skillset-in-big-demand/.

12 https://www.whatsyourbaseline.com/2019/09/the-value-of-enterprise-architecture-in-digital-disruption/

13 We will talk at length about a meta model that shows how the different architectural artifacts are "wired up" in a tool in the "Technical Governance" section.

FRAMEWORKS, METHODS, NOTATIONS, REFERENCE CONTENT — SOME CLARIFICATIONS

14 https://www.scaledagileframework.com/

15 https://en.wikipedia.org/wiki/The_Free_Software_Definition

16 https://www.omg.org/bpmn/

17 http://omg.org/spec/UML/

18 https://www.omg.org/dmn/

19 http://apqc.org/pcf

SOLUTION/PROCESS LIFECYCLE — ARCHITECTURE IS MORE THAN "DRAW IT AND FORGET IT"

20 https://www.whatsyourbaseline.com/2019/09/the-value-of-enterprise-architecture-in-digital-disruption/

21 https://en.wikipedia.org/wiki/ITIL

22 https://en.wikipedia.org/wiki/Supply_chain_operations_reference

HOW TO GET YOUR IMPLEMENTATION OFF TO A GOOD START

23 http://apqc.org/pcf

24 Reference content is discussed in chapter 4 in more depth and can be an accelerator for your program.

SELECTING AN ARCHITECTURE TOOL

25 https://www.opengroup.org/togaf

26 https://www.microsoft.com/en-us/microsoft-365/visio/flowchart-software

27 https://www.ibm.com/products/blueworkslive

28 https://aris.com/

29 https://www.mega.com/hopex-platform

30 https://sparxsystems.com/

31 Sold to IBM in 2024

32 Sold to Bizzdesign in 2025

33 This Magic Quadrant was retired in 2021, and they now offer a "Market Guide" document instead. Forrester did a similar step and offers a "Digital Process Automation Software" Wave analysis, which also covers runtime tools.

34 Go to the "Before you go" section, follow the link there and use the code to get the freebies.

BUYING AN ARCHITECTURE TOOL

35 Some additional thoughts on this could include things like extra pricing up front: i.e., how many licenses are included and how much are extra licenses, the cost of implementation fee, or an automatic yearly increase? You also might want to check on software updates — will these be included, or is that an extra fee as time goes on?

GETTING STARTED SUMMARY

36 It has the term "A practical guide..." in the subtitle for a reason :-)

DEVELOP AN ARCHITECTURE/BPM STRATEGY

37 I literally had one client saying to me, "I bought your tool, now you tell me what I can do with it". And, of course, the conversation about use cases in which she could integrate the tool in, did not resonate with her. A typical "give me an app for this, and I would rather not think" scenario.

38 Watch the commercial here: https://www.youtube.com/watch?v=-z4NS2zdrZc. And while you're at it, watch Steve's commencement speech at Stanford from 2005 as well: https://www.youtube.com/watch?v=UF8uR6Z6KLc.

39 A good concept is the Net-Map approach here: https://netmap.wordpress.com/about/.

40 https://en.wikipedia.org/wiki/Capability_Maturity_Model_Integration

41 But not the greater good, as defined the outstanding movie "Hot Fuzz" from 2007. If you haven't seen it, go spend two hours online. You won't regret it — https://en.wikipedia.org/wiki/Hot_Fuzz.

COMMUNICATE TO WIN OVER YOUR STAKEHOLDERS

42 An example of how this could look is described in Section VI "Real-world application: Business Process Management".

43 We will take a look at the enablement topic in the "Process and organizational governance" section of this book.

44 Or JIRA/Trello/Asana/... project, whatever tool you have in place for repeatable execution on a plan.

CREATE VISIBILITY INTO YOUR ARCHITECTURE (PROGRAM)

45 Again, the "meta processes" that you will define in the Process Governance area. Maybe the exercise here will give you some hints on which processes you should add to the "big map" there.

HOW TO SET UP TECHNICAL GOVERNANCE

46 The next steps are using the download client. The process in ARIS Connect is the same, but the click-through commands might be different.

47 http://notion.so/

ARCHITECTURE DATABASE SETUP AND STRUCTURE

48 In some tools, for example ARIS, you can do this in one database and use the versioning features. However, I have found this a clumsy way to do, and it requires elevated licenses to do (a "Designer" instead of a "Viewer", which costs significantly more).

49 https://www.ascm.org/corporate-solutions/standards-tools/

50 https://www.apqc.org/process-frameworks

51 http://www.opengroup.org/public/arch/p2/p2_intro.htm

52 https://en.wikipedia.org/wiki/Architecture_of_Integrated_Information_Systems

LIBRARY IMPORT AND MANAGEMENT

53 As of the time of writing this book, I am aiming to release my second book, "Successful Process Mining. A practical guide to Process Mining projects" in 2025. Depending on when you read this book, it might be out already.

54 There are tons of those out there, and one of the main differences is if they are centralized or decentralized. Take a look here for a list: https://en.wikipedia.org/wiki/List_of_version-control_software

55 And yes, I am aware that this is an ISO standard, but I have encountered (process mining) tools that rejected date stamps using a "T" to distinguish between date and time.

56 https://www.knime.com/

57 As I said earlier, this should be good practice for all kind of scripts that you will create — not only the interfaces. In ARIS this also would mean reports, semantic checks, macros, queries, etc. ... everything should move only "from left to right".

GOVERNANCE PROCESSES — THE "PROCESS OF PROCESS"

58 It might require a large sheet of paper, though ;-)

ORGANIZATIONAL GOVERNANCE DESIGN

59 For more information, read this interesting PDF: https://blog.crisp.se/wp-content/uploads/2012/11/SpotifyScaling.pdf

TRAINING AND ENABLEMENT

60 But it is also a lot of fun and I enjoyed it a lot working with my consultants, SMEs, and learning team when standing up iGrafx University in record time (3 months for the initial launch, which included a "soft launch" phase, after we had procured a Learning Management System). Everyone was super-proud when we finally went live.

61 If you are interested in my experiences when implementing a LMS (Docebo in this case; https://www.docebo.com/) please reach out. There are also open-source alternatives available, and Moodle seems to be a very popular LMS — https://moodle.com/.

62 One example of this type of software is Walk Me; https://www.walkme.com/.

63 The second screenshot is from the community forum of KNIME, an open-source data modeling tool that you might want to take a look at ... especially when it comes to Process Mining — the topic of my upcoming second book in 2025. Stay tuned and check out KNIME here: https://www.knime.com/.

64 https://asana.com/ is one potential tool for this (JIRA is another one). What you are looking for is a tool that can automate actions when triaging tickets, or when working on course updates to trigger reviews, etc.

BUSINESS ARCHITECTURE - HOW TO STRUCTURE YOUR PROCESS VIEW

65 https://en.wikipedia.org/wiki/Six_Sigma

66 https://en.wikipedia.org/wiki/Lean_Six_Sigma

IMPORT OF EXISTING CONTENT - A BLESSING OR A RECIPE FOR CHAOS?

67 https://www.opengroup.org/archimate-home

68 http://notion.so/

69 https://www.omg.org/bpmn/

70 Typically, the tracker is a product for sale, but as the reader of your book, you can get it from the freebies section of the accompanying web page to this book (info about that is in the "Before you go" chapter) — just copy the Notion page into your Notion workspace.

BASE PROCESS ANALYSIS

71 https://www.leansixsigmadefinition.com/glossary/waste/

PROCESS SIMULATION

72 https://theleanway.net/The-8-Wastes-of-Lean

73 https://en.wikipedia.org/wiki/Discrete-event_simulation

74 https://en.wikipedia.org/wiki/Monte_Carlo_method

ADDITIONAL PROCESS ANALYSIS TYPES

75 https://en.wikipedia.org/wiki/Business_intelligence

76 https://www.6sigma.us/business-process-management-articles/process-monitoring-tools-benefits-and-types/

77 https://en.wikipedia.org/wiki/Process_mining

78 https://en.wikipedia.org/wiki/Six_Sigma

79 https://en.wikipedia.org/wiki/Lean_thinking

80 https://en.wikipedia.org/wiki/Business_process_modeling

81 https://en.wikipedia.org/wiki/Process_simulation

82 https://en.wikipedia.org/wiki/Interview_(research)

83 https://en.wikipedia.org/wiki/Data_and_information_visualization

BEFORE YOU GO ...

84 This quote is attributed to many people, from the Nobel laureate Niels Bohr to the famous baseball player Yogi Berra (who made tons of other important quotes in his life). Nevertheless, it captures the problem of forecasting perfectly.

85 But I am sure that my predictions will have a longer shelf life than a head of lettuce (https://www.theguardian.com/politics/2022/oct/20/iceberg-lettuce-in-blonde-wig-outlasts-liz-truss).

86 https://www.gartner.com/en/articles/hype-cycle-for-artificial-intelligence

87 Which is a shame in my opinion. I encourage you to try GNOME out as an end user (the default desktop for popular distributions like Ubuntu or Fedora). Beautiful to look at, easy to use, and for more than a decade now easy to install.

INTRODUCTION INTO BPMN

88 https://www.omg.org/bpmn/

89 https://www.omg.org/spec/BPMN

90 https://en.wikipedia.org/wiki/Event-driven_process_chain

91 https://en.wikipedia.org/wiki/SIPOC

92 https://en.wikipedia.org/wiki/IDEF

93 https://en.wikipedia.org/wiki/Unified_Modeling_Language

94 https://en.wikipedia.org/wiki/XPDL

95 https://wfmc.org/

96 Graphic adapted from Robert Shapiro's much needed "Update on the BPMN status" from February 2010

97 https://en.wikipedia.org/wiki/Business_Process_Execution_Language

98 Please see Sebastian Stein's article about the Workflow Patterns Initiative's original 20 workflow patterns in BPMN): http://www.ariscommunity.com/users/sstein/2010-07-20-bpmn-2-workflow-patterns and http://www.workflowpatterns.com/.

99 The poster can be downloaded here: http://docenti.ing.unipi.it/m.cimino/pmi/pres/bpmn-poster.pdf/. There is also an ARIS-specific available version here: https://ariscommunity.com/system/files/cs-aris-bpmn-en.pdf.

100 Some of the objects are also used in the 1.x BPMN implementation. The listing only reflects the 2.0 implementation – e.g. a pool can be used in a 1.x BPMN process model but not in a 2.0 process model (they are only allowed in BPMN Collaboration Diagrams per the newer specification).

101 https://ariscommunity.com/system/files/cs-aris-bpmn-en.pdf

DATA MODELING

102 https://en.wikipedia.org/wiki/Conceptual_schema

103 https://www.omg.org/retail-depository/arts-odm-73/logical_data_model_concepts.htm

104 https://balsamiq.com/

CREDITS

105 https://unsplash.com/@kellysikkema

106 https://aris.com/

107 https://thenounproject.com/

PRAISE FOR "SUCCESSFUL ARCHITECTURE TOOL IMPLEMENTATION"

108 https://www.linkedin.com/in/marcmendelson1/

109 https://www.linkedin.com/in/hanneke-loefs-mos-rc-22a860/

110 https://www.linkedin.com/in/breannecasteel/

111 https://www.linkedin.com/in/martinhollingde/

112 https://www.linkedin.com/in/marcus-zschuckelt-5a1834293/

113 https://www.linkedin.com/in/michael-schank/

Made in the USA
Las Vegas, NV
01 August 2026

52152132R00207